Bean Feast

Bean Feast

An International Collection of Recipes for
Dried Beans, Peas and Lentils

VALERIE TURVEY

Illustrations by Valerie Turvey

101 Productions
San Francisco

Printed and bound in the United States of America.

Published by 101 Productions
834 Mission Street
San Francisco, California 94103

Distributed to the book trade in the United States by
Charles Scribner's Sons, New York.

First published in Great Britain 1979
by Pitman Publishing Limited
39 Parker Street, London WC2B 5PB

Library of Congress Cataloging in Publication Data

Turvey, Valerie.
    Bean feast.

    Includes index.
    1. Cookery (Beans)   I. Title.
TX803.B4T87   1979   641.6'5'65   79-5447
ISBN 0-89286-158-4

Contents

Beans, lentils and other dried legumes first began to feature in our family diet as an alternative protein source when meat prices began spiralling – now they are popular simply because they taste good. Now that the days of cheap food have finally vanished into the memory, the only economies that any housewife can make are in the kitchen rather than in the shops, and this means cooking in a way that our grandmothers took for granted. Simple, basic foods, fresh or dried, cooked with love and care and with a pleasant 'homely' feel to them, are surely better than anything poured out of a can and reheated. Although the press and advertising media like to depict us all as a breed of convenience food consumers, can-opening, package orientated and gullible to the glossy advertisements, there is an ever-growing number of people like myself who, whether from necessity or pleasure, or in my case both, prefer to take the time and trouble to produce real 'home cooking' from good, honest ingredients.

As an art student in the early sixties with a longer bill for paint than for food, I became very aware of the need to find the cheapest and most nutritious food available. At the same time I needed to satisfy my taste for exotic and unusual foods, which was not really helped by the expensive ingredients in all the coffee-table cookbooks. I only wish that the numerous wholefood shops that have mushroomed recently in many towns and cities had been in existence then. It was primarily the opening of our local wholefood shop, 'The Beggar's Banquet', that prompted me to write this book. Visiting the shop with a health food addict friend, I asked him how he cooked aduki beans. His answer was 'Boiled, of course.' I think it is lack of knowledge about these excellent basic foods that leads many people in this country to stick to the split pea soups and boiled butter beans they know already. By using recipes from nations that use these dried legumes more imaginatively and with greater necessity than we do at present, I hope this book will give its readers the confidence and enthusiasm to try cooking them in more interesting ways.

The wholefood shops themselves contain a bonus in sights and smells that you will never find in a supermarket – sacks of grains, beans, fruits and nuts, herbs and spices in a bewildering variety of earthy colours and rich textures. Some of my earliest memories are of delicious smells and fascinating assortment of jars, boxes and sacks of goods piled and stacked into my great-aunt's shop. Everything from dried fruit to rice, split peas to biscuits, was sold loose, weighed on my great-grandfather's brass scales, scooped into thick blue paper bags or wrapped neatly in greaseproof paper. Hams and legs of pork, tied in muslin, floated mysteriously in great stone brine baths in the back yard; slabs of dark bacon rested on the red and chrome slicer, and enormous wedges of Cheddar cheese waited under a cover to be sliced and sold with one of the jars of home-made chutneys and pickles. Above all there was the all-pervading warm and friendly smell of real honest-to-goodness food, all long-since bulldozed away to make room for a supermarket which, however colourful, hygienic and tuneful, will hardly fill our children's memories with such pleasant images. Because of this I especially welcome these little wholefood shops, where no one complains if my small son rummages in the big bins of black eyed peas or plays with scoops of lentils from the sack: incidentally, lentils and beans make excellent play material for collages, playing shops, loads for dumper trucks, etc., as long as they are kept away from smaller children who may try the age-old trick of inserting them in ears and noses.

History of Beans

Beans and lentils have provided man with a basic source of protein for thousands of years. The Aztecs, Egyptians and the ancient Chinese all cultivated various legumes and stored their dried seeds for winter food, to supplement a diet of grain and fruit. Chick-peas, fava beans and lentils were grown in the Middle East as early as 8000 BC, while the lima bean and its relatives, including early forms of haricot, were cultivated on the high plains of Peru from circa 6000 BC. The Chinese started cultivating the soya bean around 3000 BC and, with their usual ingenuity, produced a wide range of nutritious vegetarian products over the years, including soya milk, bean curd and soya sauce. Haricot beans were brought to Europe from the Americas by contemporaries of Columbus, whilst even earlier chick-peas were introduced into Spain from North Africa. Beans were popular in Ancient Greece, although the Romans considered them unlucky. Perhaps this was because they were used as a means of voting in elections, white being for and black against, rather than having anything to do with their edible qualities. The Romans preferred chick-peas, thus demoting peas and beans to the poorer classes – a position they have held throughout history.

In many cultures, superstition connects beans with ghosts and witches, death and magic – even the story of Jack and the Beanstalk relies on their 'magical' qualities. Bean feasts, or Beanos, were held in most Celtic regions, and on Twelfth Night beans were used in the festive cake to denote the King for the evening's revels – recipes for these cakes are still used in some European countries, the Galette de Rois of France being one example.

Today, legumes are the staple diet in many of the poorer nations and less affluent communities around the world, especially in South America, the West Indies, India and the Middle East. In northern Europe, the most commonly used legumes are split yellow and green peas, haricot beans and lentils, usually served as thick winter soups and purées, rich with ham stock, pork hocks, smoky bacon and coarsely cut sausages. France and Italy have produced many classic dishes using haricot beans, marrying them with highly-flavoured pork and goose drippings in cassoulet, or with olive oil, vegetables and tuna in salads and thick vegetable stews. In Spain, the chick-pea is still the most popular legume, as it is in North Africa where, with ful beans and lentils, it is flavoured with lemon, olive oil and garlic. Greece has its own particular favourite in the dried broad bean, probably one of the best flavoured of all legumes. In India lentils, under their Hindi name dal (or dahl), are eaten as the main source of protein in most of the vegetarian south. The variety of methods they use for preparing lentils for breads, fritters, salads, pancakes and vegetable dishes can only be matched by the soya bean products of China and Japan. In South America, rice and beans, spiced and fried, constitute the daily meal in many poor communities, as they do in the West Indies. Britain seems to have adopted the two least interesting varieties – the butter bean, with its memories of school meals, when it was served either tough or overcooked with a ladle of cooking water, and the small pink lentils used to thicken soups. Both of these are used with far more ingenuity, and taste, elsewhere in the world, and the Indian and Middle Eastern methods of serving lentils makes them barely recognisable as the same ingredient which goes into those mushy pink soups.

Beans are Good for You!

The nutritional qualities of beans and lentils are already well expounded and it would be out of place, as a layman, for me to attempt to compete with the many excellent books on nutrition on the market. Briefly, legumes are an excellent source of incomplete protein: that is, they do not contain all the essential amino acids needed to maintain growth. However, if used in conjunction with other protein sources, either vegetable (rice or other grains, including wheat products such as bread and pasta) or animal, they can become complete. All legumes contain Vitamin B in most of its forms and a few contain Vitamin A. Lentils, in particular, are rich in iron and contain some calcium. However, because they lack Vitamin C it is wise to complement bean dishes with either a fresh salad, lightly cooked vegetable dish or raw fruit, depending on your menu.

Although these recipes are not all vegetarian, anyone who requires convincing of the need to eat less meat and more vegetable protein for the sake of both their health and pocket would do well to read *Diet for a Small Planet* by Frances Moore Lappé, published by Ballantine, and the complementary *Recipes for a Small Planet* by Ellen Buchman Ewald, by the same publisher. Their research into the ability of vegetable proteins to complement each other and their stress on natural rather than processed foods makes compelling and convincing reading – heartening news indeed to find that meat and fish, foods which have almost priced themselves out of our pockets, are not as essential as we have been brought up to believe.

Buying Beans

Apart from their comparative cheapness, beans and legumes have excellent storage properties, infinite variety, and are virtually foolproof to cook. The long cooking times of some of the larger and harder beans may seem an extravagant use of fuel, but comparing the price per pound of any beans to even the toughest stewing meat, the difference is staggering, and the cooking time much the same. Select a shop with a high turnover of dried foods – the wholefood shops already mentioned, health food shops and the Indian, Chinese and West Indian grocers catering for immigrant families in the larger cities have excellent varieties at a competitive price. Avoid, if possible, packaged, and especially fancily-wrapped, varieties – aduki beans in a wholefood shop cost a fraction of the price of a packaged variety in a health food shop in the same city. Avoid also the packets tucked away at the back of the shelf in grocers and supermarkets, as even the longest cooking will not resurrect elderly beans. Buy small quantities of new varieties, and always store beans in airtight containers. Their decorative qualities, when stored in glass, are obvious and bring a splash of warm, earthy colour to the kitchen.

Anyone who has room to grow green beans might like to try growing their own haricot beans. The most readily available varieties are Chevrier Vert and Comtesse de Chambord, both of which can be eaten fresh, either in the pod or as flageolets. For haricots, they are left to dry on the plant until the pod and seed are white.

Sprouting Legumes

Most legumes can be sprouted, although the best known are the mung bean sprouts used extensively in Chinese cooking. Soya and aduki beans and urd dal are also sprouted and served raw or lightly cooked in salads, vegetable dishes, egg dishes and soups. Although I have tried all the recommended methods the only one I find reliable is as follows, remembering that constant drainage and frequent fresh water are the essential ingredients.

Soak the beans or lentils overnight. Drain and place in a colander which has been lined with damp muslin (or towel or tea-towel). Cover with a similar piece of damp cloth, folded into three or four layers. Water through the material 3 or 4 times a day for 4 to 6 days. Keep in a warm room, 65°F (18°C) is hot enough. Check after four days, discarding any legumes that have not sprouted. Do not let them continue sprouting after seven days, and remember that infrequent watering can cause mould. They can be stored in the refrigerator for up to 36 hours but should, ideally, be eaten immediately. They are not only cheaper than the tinned Chinese bean sprouts but the difference in taste and texture has to be tried to be believed. I have given a few basic recipes for using them but suggest you refer to one of the numerous, excellent books on Chinese cooking. One cup (250ml) lightly packed bean sprouts equals 2 ounces (50g).

Approximate sprouting times:
Aduki beans	– 4 days.
Soya beans	– 5 days.
Mung beans	– 5 days.
Urd dal	– 6 days.

General Cooking Instructions for Legumes

Always wash legumes thoroughly, picking over and discarding any stones or seeds.

Individual cooking times are given in the glossary and on page 11. Two basic cooking methods can be used, each having their own advantages in time and convenience, but there is little difference in the end result.

1. **Long:** Soak overnight in plenty of cold water. Drain and cook in unsalted water to cover.
2. **Short:** Put beans in a saucepan. Cover with 2 inches (5cm) unsalted water and bring to the boil. Cook for 2 minutes. Remove from the heat and soak for 1 hour. Continue cooking in the same water.

In general lentils and split peas do not need soaking, unless they are to be ground in the raw state.

Salt is best added at the end of the cooking time, especially in the case of Great Northern and soya beans, as salt toughens their outer skin.

Soaking and Cooking Times

Be careful not to soak the beans over-long, as they may start to ferment.

The cooking times given here should only be used as a guide, as the freshness of the beans will affect the timings (the fresher the bean, the less cooking it will require).

TYPE	SOAK-ING	COOK-ING	PRES-SURE COOKER (soaked)	PRESSURE COOKER (unsoaked)
Aduki	long	2 hours	20 mins	30 mins
Black beans	long	2 hours	20 mins	25 mins
Black eyed peas	short	45 mins	15 mins	20 mins
Butter beans	long	45–60 mins	20 mins	25–30 mins
Chick-peas	long or short	1 hour	15 mins	35 mins
Dals	none	30 mins	10 mins	10 mins
Fava beans	24 hours	3 hours	40 mins	1 hour
Ful nabed	24 hours	3 hours	40 mins	1 hour
Great Northern beans	long	1 hour	20 mins	25 mins
Lentils, brown	none	40 mins	15 mins	15 mins
Lentils, green	none	40 mins	15 mins	15 mins
Lentils, red	none	30 mins	10 mins	10 mins
Mung beans	1 hour	45 mins	12 mins	20 mins
Peas, split	none	30 mins	10 mins	10 mins
Peas, whole	long	40 mins	15 mins	20 mins
Pigeon peas	none	30 mins	10 mins	10 mins
Red kidney beans	long	1 hour	20 mins	25 mins
Small white beans	long	2 hours	20 mins	25 mins
Soya beans	long	2 hours	25 mins	30 mins

Some Notes on Using the Recipes in this Book

When soaking times are not given in individual recipes, use the table on this page as a guide. Beans and lentils should always be drained after soaking and before use unless the recipe states otherwise.

All the recipes that follow can be adapted for use with a pressure cooker which, although not essential, certainly takes some of the sweat out of cooking dried foods by reducing their cooking time substantially. It also eliminates the need for soaking all but the toughest beans. Most of the recipes require the beans to be cooked, either alone or with flavourings, before combining with other ingredients, and I find the pressure cooker ideal for this preliminary cooking. Always refer to your pressure cooker instructions for exact times, although the table on this page can be used as a general guide.

Although a blender is also not essential, it takes the time and effort out of puréeing soups and, in particular, grinding dry beans and lentils for flours and batters. However, take care not to overblend soups or they will end up with an excessively smooth, uninteresting texture.

Blanched bacon Simmer sliced bacon for 10 minutes in 4 to 6 cups (1 to 1½ litres) water, then rinse and dry on paper towels. Fresh side pork can be substituted, if available.

Chillies are a matter of taste, and amounts can always be increased or reduced. If in doubt, start by adding less. If whole chillies are used, they can be removed during the cooking to lessen their flavour.

Ghee is clarified butter used in Indian cooking; regular butter or oil can be substituted.

Herbs should be fresh or, if dried, bought loose in small quantities.

Oil is the best you can afford – I use soya oil for general cooking, saving olive oil for salads and those recipes which need it for its flavour as well as its cooking properties.

Paprika can also be altered to suit personal tastes.

Pepper means freshly ground black pepper.

Peppers, green and red can be peeled by placing under a hot broiler until they become charred. The skins split and are easily removed.

Salt means sea salt, whenever possible.

Spices are best when freshly ground, and are cheaper if bought loose from wholefood and health food shops or Indian grocers.

Stock should be home-made if at all possible.

Tahina paste is made of ground sesame seeds, and is available in health food stores and Middle Eastern markets.

Vinegar means wine or cider vinegar.

Metric equivalents are approximate, and have been adapted for each individual recipe.

Unless otherwise stated, all recipes are for 4 to 6 people (depending on appetite and whether the dish is served on its own or not).

Glossary

Aduki Beans (Azuki, Adzuki, Adsuki)

A small, oval bean ¼ inch (½cm) long, with a dull red seedcoat and thin white keel. Grown and eaten in China and Japan since prehistoric times, the beans grow on small bushy plants and are eaten fresh, dried, sprouted or ground into a flour. In Japan, rice is steamed with aduki beans to stain it pink to make Sekihan, a festive dish served at weddings (p. 155). Some Japanese cakes and confections are also made from aduki beans.

These beans are good simply boiled and served with butter or in salads. The seedcoat is quite tough and they need fairly long soaking and a two hour cooking. Aduki beans contain 25% protein and have a high mineral content plus iron, calcium and thiamine. Available from health food shops, but cheaper (less than half price) if bought loose from wholefood shops.

Black Beans (Frijoles Negros, Turtle Beans)

Large, narrow kidney bean, ¾ inch (1½cm) long – charcoal black with a white keel. A native of South America where it is a staple food – boiled, spiced and fried. They appear daily in most poor communities and as Feijoada Completa (p. 74) constitute the national dish of Brazil. In the West Indies and the Caribbean they are eaten with rice as a major source of protein – the colour combination of white rice and black beans gives the Cuban dish of Moros y Cristianos (Moors and Christians) its name (p. 124).

They can be soaked by either method and take approximately 1 hour's cooking. One of my personal favourites, black beans have a distinctive taste and good texture, combining particularly well with garlic, chilli and tomatoes. Nutritionally similar to haricot beans, they are available at most delicatessens, wholefood and health food shops, West Indian grocers and some supermarkets.

Black Eyed Peas (Black Eyed Beans, Cow Peas, Oea Beans, China Beans, Marble Beans)

A cream kidney-shaped bean, ½ inch (1¼cm) long with a purple-black keel. A native of China but mostly associated with the cooking, literature and folklore of the deep south of America, where they were introduced in the eighteenth century. They are also used in Indian cooking (LOBIA DAL) and African dishes. They are possibly best known in Hoppin' John, a 'soul' dish from the deep south of the United States which consists of rice and black eyed peas served with a variety of pig extremities (p. 124).

A short soaking and cooking time is required and after 30 minutes' cooking they need checking as they can disintegrate. Black eyed beans are best in salads, spiced or with yoghurt, and make excellent fritters in the African and Brazilian tradition. They contain about 22% protein, some iron, calcium, Vitamin B complex and, unlike most beans, some Vitamin A. Widely available from larger stores, health food stores, wholefood shops and delicatessens.

Broad Beans, see Fava Beans

Brown Beans (Swedish Beans)

Small, oval brown beans, ¾ inch (1cm) long. Used in sweetened purées to be served with ham or pork. Not widely available, but small white beans can be substituted.

Butter Beans (Large Limas)

A large, creamy white bean 1¼ inch (3cm) long, part of the haricot family. Coarser in both flavour and texture than haricots, they are also available canned. At their best they make a passable substitute for haricots in salads, purées and soups if other flavourings are present to make them more interesting.

They need long soaking and careful cooking, as once they soften they become mushy very quickly. Because of their all-pervading cooking odour, which has given them a bad name, do not drain into the sink. Add salt towards the end of cooking and if you have time, and patience, skin before serving. Widely available at grocers, supermarkets, etc.

Channa Dal (Chenna, Chanai, Arhur Dal, Toor Dal, Tur, Toer, Toovar Dal, Ooloonthoo, Pigeon Peas)

A small, split yellow pea, ¼ inch (½cm) round, rather soft and quickly cooked. Usually used in meat or vegetable dishes or in purées. A very good substitute for chick-peas if making Bessan (p. 142) at home, as they are less difficult to grind.

The word channa on its own is used in its literal translation of gram, meaning legume. This can be confusing in some recipes where, like the word dal, channa is used as a loose umbrella term. Kabli channa is another name for chick-peas. It is sometimes available packed in oil, when it is known as TOER or TUR DAL. Yellow split peas can be substituted in mixed vegetable or puréed dishes as the texture is identical, although the flavour is not the same. Widely available in delicatessens, wholefood stores, Indian grocers.

Chick-Peas (Garbanzos, Ceci, Hummus, Kabli Channa)

Round, beige-yellow peas, ⅜ inch (1cm) in diameter, with a pronounced beak-like sprout. Impossible to overcook and one of the best, in taste, texture and versatility, of all legumes. Popular in most Mediterranean countries, they originated in Africa and were introduced into Europe by the Carthaginians when they entered Spain where, as GARBANZOS, they still play an important role in regional cooking. In Italy, as CECI, they are used in soups, with pasta, breads, batters, fritters, pastries and pancakes. They can be sprouted and eaten fresh.

Chick-peas are very hard, but can usually be cooked after only 3 hours' soaking. However, if you are not sure of their age, soak overnight.

Dal (Dhal)

Generic Hindi name for legumes and lentils (see individual dal). Widely used in Indian cooking. Little or no soaking, a good source of iron and some Vitamin B.

Fava Beans (Broad Beans, Ful Medames, Ful, Ful Misri)

Eaten fresh in England and Europe, when dried the fava is a small, hard, brown broad bean. It is used extensively in the Middle East, especially in Egypt. It has a tough leathery, tan-coloured seedcoat with a darker brown keel and is ¾ inch (2cm) long.

These beans need very long soaking, depending on their age – in some cases as much as 36 hours – and approximately 1 hour's cooking, again depending on age. They have an excellent flavour and texture, although some people may prefer to skin them after cooking. Good, whole or puréed, in salads with olive oil and lemon juice. Available from wholefood stores and ethnic markets.

Ful Nabed

The white counterpart of ful medames. A dried, creamy-white broad bean, approximately ¾ inch (2cm) long, with a tough seedcoat, they need long soaking and cooking. Used in the Middle East and Greece, mostly in salads and soups, they are especially popular as Ta'amia (crushed skinned beans, deep fried in oil – p. 147) and Kushari (beans, pasta and fried onions), both of which are sold by Egyptian street vendors. Buy from similar sources as ful or fava beans.

Garbanzos, see Chick-peas

Great Northern Beans, see Haricot Beans

Haricot Beans (Great Northern Beans)

Many varieties and colours exist, including red, but the most commonly sold are creamy white, ⅜ inch (1cm) long, slightly kidney-shaped with a tough seedcoat. Haricot beans are the dried, mature white seeds of the green bean, which is also known as the haricot vert or French bean.

Originally a native of the Americas, beans from the haricot family were brought to Europe in the 1520's via Italy and are now possibly the most widely known and used of all beans. They became especially popular in Tuscany, where most of the best haricot dishes come from, and even today Tuscans are known as *mangiafagioli* (bean eaters). Of their many dishes the most famous is probably Fagioli nel fiasco (beans cooked in a chianti bottle), but the most delicious is Fagioli Toscanelli con Tonno (p. 57) – a classic salad of beans and tuna in olive oil and vinegar. In France, haricot beans are synonymous with Cassoulet (p. 73), or the thick bacon and bean stews of the south-west, or Gigot à la Bretonne (p. 84) in the north.

Haricot beans are best bought in autumn and, if possible, loose from delicatessens, wholefood stores, etc., as some supermarket packets may be very old. If you can find the superior French soissons or the Italian cannellini, they are worth the extra cost. Larger varieties include Comtesse de Chambord and Burma. The most common variety in the United States is the Great Northern. Haricot beans contain 23% protein and iron, calcium and Vitamin B.

Hummus, see Chick-peas

Lablab Beans

A nutritious brown bean with a white keel. A native of India, it is now cultivated and used in Egypt and the Middle East. Rarely imported to other countries.

Lentils

The seeds of a small shrub family found in the Mediterranean area and cultivated since about 8000 BC. Eaten by the ancient Egyptians and Romans and also in medieval times, they have formed a nutritious and comforting base for many family meals through the ages, as they still do today in many poor communities. Although mostly used in soups in Europe, they are treated more imaginatively in the Middle East. There are many varieties of lentil, including the numerous types of dal in India, but the main kinds available elsewhere are red, green and brown.

Lentils need no soaking and, unless needed for a purée, must be cooked carefully as they soften quickly. They contain 25% protein and are rich in iron and Vitamin B. The better green and brown lentils from France are usually only obtainable at delicatessens, but the red variety is widely available in grocers, health food stores, etc.

Lentils, Brown

Slatey-brown, shiny and flat, these are $\frac{1}{4}$ inch ($\frac{1}{2}$cm) round. The *lentilles blondes* of France, which is where the best ones come from, they are the most flavoursome of the three lentils. Excellent served *al dente* with butter, they are also good in Middle Eastern dishes where they combine well with olive oil and rice. They need careful cooking and should be checked after about 20 minutes.

The BROWN MASOOR LENTIL, which is the red lentil in its seedcoat, cannot replace the brown lentil in European recipes but can be used in Indian and Middle Eastern dishes.

Lentils, Green

Larger than the red lentil, they have a dull olive colour, are quite flat and $\frac{1}{4}$ inch ($\frac{1}{2}$cm) round. The best French ones, from Le Puy, are hard to come by, but all the French varieties are better than average. Firmer than red lentils, they take about 20 to 30 minutes to cook, depending on age and variety.

Lentils, Red (Egyptian Lentils, Ads Majroosh)

A flat salmon-pink lentil, $\frac{1}{4}$ ($\frac{1}{2}$cm) round, with a higher protein content than other varieties. The chief problem is that overcooking can make them mushy, so if you want to serve them whole, start checking after 10 minutes. They can be ground with rice. Available from health food shops and Indian grocers, and can also be sprouted.

Lima Beans (Baby Limas)

A creamy white, kidney-shaped bean, part of the haricot family which has been cultivated in South America since 7000 BC. This floury bean has a very good flavour and is popular in the Americas, especially Brazil.

Lobia Dal, see Black Eyed Peas

Masur Dal (Malika Masoor, Masoor Dal, Mysore Dal)

A small flattish lentil, $\frac{1}{4}$ inch ($\frac{1}{2}$cm) round, salmon-pink in colour, which is virtually identical to the common red lentil of Europe. When the brown seedcoat is left on it is known as BROWN MASUR and can be used in Middle Eastern and Indian recipes specifying brown lentils, but it cannot be substituted for the French brown lentil.

Masur dal cooks quicker than other lentils and, in its husked state particularly, rapidly disintegrates. This is the lentil to use in recipes which state 'dal' and leave you wondering which one. Best suited to purées, soups and khichiris (p. 130).

Motth

A brown Indian lentil, similar to masur dal, for which it can be substituted.

Mung Beans (Moong Dal, Mung Dal, Green Gram, Golden Gram, Lou Teou)

A small $\frac{1}{4}$ inch ($\frac{1}{2}$cm) oval bean, yellow with an olive green seedcoat and thin keel, which is sold either whole or husked. They are perhaps best known in their sprouted form as the tasty BEAN SPROUTS in Chinese cooking, but in India they are also eaten whole with spice or butter, puréed, or ground to a flour. There is a brown and a black variety, although neither of these is so readily available.

They can be soaked for 1 hour before use but this is not essential. Mung beans contain 24% protein. Obtainable from Chinese and Indian grocers, wholefood and health food shops.

Navy Beans, see Small White Beans

Peas, Blue

A dark bluish-green whole dried pea, firmer and with a better flavour than marrowfats. Cook and use as marrowfats (see under Peas, Green, Whole). Available from wholefood shops and some grocers.

Peas, Green, Whole

The dried marrowfat garden pea, dull green and $\frac{3}{8}$ inch (1cm) in diameter, with a wrinkled skin. Used mostly in northern Europe as a nourishing winter standby in soups and thick purées. Dried peas should not be cooked with baking soda. While the addition of baking soda will speed up the cooking time, it will also make the peas mushy and will dull their flavour. The use of baking soda should be avoided in cooking all legumes and vegetables as it destroys valuable nutrients and breaks down the cellular structure. Soak the dried peas overnight and then cook them in plenty of water until *al dente* (if such a term can be applied to dried peas) and they will make an interesting winter vegetable.

In soups and ribsticking winter stews they combine well with bacon, ham and smoked sausage. They contain iron, protein, calcium and some Vitamin A and B.

Peas, Green, Split

$\frac{1}{4}$ inch ($\frac{1}{2}$cm) in diameter, these husked and split peas are used mostly in soups. Dried peas have a long history and were used in Egypt and Ancient Greece, but fell out of fashion under the Romans, although they were still eaten by the poorer communities. Dried peas and pea soups are still found as good, cheap, filling fare in rural cooking today. They need long soaking and cooking, and are widely available.

Peas, Yellow, Split

Dried yellow peas, husked and split, $\frac{3}{8}$ inch (1cm) in diameter. They need no soaking and can be cooked in 30 minutes, or slightly longer for a purée, which suits their rather mushy texture.

Yellow split peas are mostly used in northern Europe as a winter vegetable and in soups. Although larger than channa dal and pigeon peas, they can be used if these are not available. Most commonly used in Britain to make Pease Pudding (p. 116) or pea soup with bacon and vegetables. The Germans and Scandinavians make excellent use of split peas in their thick winter soups, using pork, smoked bacon, ham and sausages to complement the peas. Widely available at grocers.

Pigeon Peas (Gandules, Red Gram, Congo Peas, Gungoo Beans)

A small, yellow, soft-textured pea, $\frac{1}{4}$ inch ($\frac{1}{2}$cm) in diameter, with grey markings. It is known as CHANNA DAL when sold in the husked, split state. Of African origin and grown by the Egyptians since 2000 BC. Used in African and West Indian cooking, spiced and puréed. Fairly widely available from health food stores and West Indian grocers.

Pink Beans

A dull pink version of the red kidney bean, used in South America. If unavailable, substitute red kidney beans or pinto beans.

Pinto Beans

A small oval kidney bean from South America, $\frac{3}{8}$ inch (1cm) long, and beige in colour with tan blotches. The name comes from the Spanish *pinto*, meaning painted. They are widely available, but if unobtainable, substitute red kidney beans.

Rajma Dal, see Red Kidney Beans

Rangoon Beans

A red haricot bean, not widely available.

Red Beans

A small, dark-red oval bean that can be substituted for red kidney beans.

Red Kidney Beans (Rajma Dal, Habichuelas)

A large-shaped bean, $\frac{3}{4}$ inch (1$\frac{1}{2}$cm) long, dark red in colour, becoming tan when cooked. Native to South America, where it is a staple food, it is also found in India.

Red kidney beans were eaten simply boiled by the South American Indians before the Conquistadors introduced pigs to the continent, but with the lard which then became available a series of dishes evolved including Refried Beans (p. 98), stews and chilli-flavoured beans. Red kidney beans were also the staple food of the American cowboy and are a debatable ingredient in Chilli con Carne (p. 78) – a Texan, not Mexican, dish. Texans claim the genuine chilli contains no beans, but red kidney beans are usually included by other Americans. In Russia they make salads from red beans and dress them with sweet and sour dressings based on damson plums (p. 60) – an excellent combination.

Either method of soaking is applicable and 1 hour's cooking is usually sufficient. They have a good texture and flavour, and combine well with tomatoes, spices and pork. Widely available in larger stores, wholefood shops and delicatessens.

Red Kidney Beans, Dark

A large bean that retains its dark red color when cooked. Often used in salads.

Saluggia Beans

An oval pink kidney-shaped bean, ¾ inch (1½cm) long, with tan patches. Used in Italian and South American cooking. If unavailable, substitute red kidney beans or pink beans.

Sieva Beans

A small lima bean – substitute haricot beans if unobtainable.

Small White Beans
(Navy Beans, Pea Beans)

A small white oval bean that requires long cooking. Used for baked beans; can be substituted for haricot beans.

Soya Beans

A small yellow-beige bean, ¼ inch (½cm) in diameter, round and very hard, becoming oval when soaked. It is a native of China, where they have been cultivated since 3000 BC as a source of cheap protein. Over thirty varieties exist in many colours, although only a few varieties are commonly available.

Although introduced into Europe in the seventeenth century, the bean was not fully exploited, except as cattle fodder in the United States, until comparatively recently. Now, however, its importance as a high-protein, low-starch food has led to an almost cult-like position in the vegetable world, and it is even made to look like meat, for which it is sold as a substitute – rather a waste of processing, as it is a perfectly good vegetable in its own right.

In China and Japan the soya bean is really used to the full and there are enough soy products to fill a book – none of which masquerade as meat. Soy sauce is found in many thicknesses and depths of flavour from the delicate Japanese Usukuchi to the thick soy jam. Fermented black and brown soya beans have the same salty pungency as olives and anchovies. Japanese Tofu and Chinese Bean Curd (p. 141) are made by grinding and straining the beans to produce this cheese-like product and the stronger bean curd cheese. Japanese miso (fermented bean paste) in its many forms – shiro-miso (white), aka-miso (red), hacho (the most nutritious), kome, mugi, sendai – has a salty-sweet taste which makes it an excellent flavouring for vegetable soups, spreads and salad dressings – as well as in traditional Japanese cooking. Soya oil, soya milk and fermented bean curd skin are other derivatives. Soya beans are also eaten fresh and sprouted, when they produce a large, crisp bean sprout, a little coarser in taste than the mung bean sprout.

Soya beans have a good meaty texture which retains a distinct 'bite', and they require fairly long soaking and cooking – ideal pressure cooker food. Their main disadvantage is the tendency to froth in the saucepan and produce a scum, which then has to be removed. Soya beans absorb a lot of water, and as they swell considerably when soaked, a few go a long way. Personally, I find their flavour less interesting than black beans or chick-peas, and prefer to use them in spiced dishes or with garlic, herbs and tomatoes. They can be used for most of the haricot bean dishes, except the classic French and Italian ones.

With a protein content of 34%, soya beans remain possibly the cheapest protein source available. Widely stocked in health food stores, wholefood shops and Chinese supermarkets.

Tepari Beans

A Mexican haricot bean especially resistant to heat and drought. Substitute haricot beans.

Urd (Urad) Dal (Maahn, Black Gram)

A small round lentil, $\frac{1}{4}$ inch ($\frac{1}{2}$cm) in diameter. When sold with its black seedcoat intact it is known as BLACK URD, but when husked it is creamy yellow and known as WHITE URD. Fairly soft, it has a more distinctive taste than channa dal or masur dal.

It combines well with butter, spices and other lentils, and is often served puréed. Urd dal can be used in batters for cakes and pancakes, when it is substituted for masur dal in Indian recipes. Known as ADS MAJROOSH to the Arabs, who use them in a variety of ways.

soups

Black Bean Soup

1 cup (250ml) black beans, soaked
4 cups (1 litre) water
1 onion, chopped
1 stick celery, chopped, including leaves
½ pound (255g) pork hocks or neck bones
1 bay leaf
Salt and pepper
1 tablespoon cider vinegar
Garnish:
1 lemon, sliced
2 hard-cooked eggs, chopped
Chopped parsley

Simmer the beans in the water with the onion, celery, pork and bay leaf for 2 hours. Remove the pork and bay leaf and skim off any scum from the surface.

Purée or mash the beans and return to the soup. Season with salt, pepper and vinegar. Bring to a simmer and serve topped with the lemon, eggs and parsley.

Portuguese Bean Soup
Portugal

1 onion, finely chopped
1 small green or red pepper, finely chopped
2 tablespoons olive oil
1 cup (250ml) black eyed peas, soaked
½ teaspoon paprika
Salt and pepper
1 tablespoon lemon juice
4 rounds of toast

Cook the onion and pepper in the oil for 10 minutes until soft but not brown. Add the beans, the paprika, salt and pepper and enough water to cover. Bring to the boil, cover and reduce the heat. Simmer for 45 minutes to 1 hour. Stir in the lemon juice.

Strain the beans, reserving them for another meal, place the rounds of toast in a tureen or in individual bowls and pour the soup over. Alternatively you can purée the soup, although this is not an authentic way of serving it.

Cuban Black Bean Soup   *Cuba*

This soup has an excellent flavour although its colour is not as appetizing as some.

1 cup (250ml) black beans, soaked
2 cups (500ml) chicken stock
1 clove garlic
1 onion, finely chopped
¼ pound (100g) ham, chopped
1 tablespoon oil
1 large tomato, peeled and chopped
Pinch ground cumin
Salt and pepper
1 tablespoon vinegar

Cook the beans in the chicken stock for 1 hour. Drain and reserve the liquid. Mash the beans or purée coarsely.

Cook the garlic and onion in a little oil until soft, add the ham, tomato and cumin, stirring until the mixture thickens.

Combine the bean purée, the cooking liquid and the tomato mixture. Bring to the boil and season with salt, pepper and vinegar. Serve hot.

Ärter med Fläsk   *Sweden*

The traditional Swedish Thursday night supper, followed by pancakes and jam. A very thick, ribsticking stew designed to keep out the cold. The pork can be served separately after the soup.

1½ cups (375ml) split yellow peas
¾ pound (350g) salt pork or fresh side pork
1 onion studded with 3 cloves.
1 onion, chopped
3 cups (750ml) water
½ teaspoon dried marjoram
½ teaspoon dried thyme
Salt

Boil the peas in the water for 1 hour, adding the pork, onion, chopped onion and herbs as the water comes to the boil. The peas should be tender but not mushy.

Remove the whole onion and the salt pork. Slice the salt pork and put one slice in each bowl. Season the soup and pour over the salt pork.

Pea Soup

A traditional pea soup which is always popular in my family. It is still very economical, especially if you can get the all-important ham bone for the asking. It is a very filling soup and also one which can stick as it thickens, so use a heavy-bottomed pan.

1 cup (250ml) split yellow peas
2 carrots, peeled and chopped
2 onions, chopped
4 potatoes, peeled and chopped
2 leeks, chopped
1½ cups (375ml) water
1 ham bone
1 teaspoon mixed dried herbs
¼ pound (100g) bacon pieces, or small piece of slab
  bacon
Salt and pepper
Dumplings:
⅔ cup (150ml) all-purpose flour
2 ounces (50g) shredded suet or 4 tablespoons lard
½ teaspoon salt
1 tablespoon chopped parsley
4 tablespoons water

Put all the soup ingredients into a large saucepan and bring to the boil. Cover and simmer for 1½ to 2 hours until the soup is thick and the peas and potatoes have partly disintegrated.

To make the dumplings, mix the flour, suet, salt and parsley with enough cold water to make a soft paste. Bring the soup back to the boil and drop heaping tablespoonfuls of dumpling paste into the soup. Cover and cook over a moderate heat for 20 minutes.

Remove the ham bone and serve the soup as a main course. If a bacon piece is used, remove it, cut it into coarse chunks and return the meat to the soup.

Middle Eastern Pea Soup   *Middle East*

A totally different soup, although using split yellow peas, lighter and lemony and typically North African.

1 cup (250ml) split yellow peas
1 stick celery
4 cups (1 litre) chicken stock
Salt and pepper
1 teaspoon ground cumin
Juice of 1 lemon
Garnish:
Chopped parsley

Cook the peas and celery in the stock for 1 hour. Purée. Season to taste with salt, pepper, cumin and lemon juice. Reheat and serve sprinkled with parsley.

Gule Aerter   *Denmark*

Probably Denmark's favourite soup – thick and savoury with frankfurters and ham.

1 cup (250ml) split yellow peas
3 cups (750ml) water
1 teaspoon salt
¾ pound (350g) pork butt
1 onion, chopped
2 carrots, peeled and sliced
2 large potatoes, peeled and chopped
3 leeks, chopped
2 sticks celery, chopped
½ pound (225g) ham, cut in strips
¼ pound (100g) frankfurters

Cook the peas in the salted water for 45 minutes to 1 hour or until very soft. Blend or sieve to give a smooth purée.

Meanwhile, cook the pork butt in enough water to just cover for 1 hour or until tender. Add the vegetables and simmer for 20 minutes or until the vegetables are just tender.

Remove the pork butt and cut into slices. Add the puréed peas to the vegetables, together with the ham and frankfurters. Bring almost to the boil and cook for just 10 minutes. Remove the sausages and ham.

Serve the soup without any accompaniments, followed by the sliced pork, ham and frankfurters, to be eaten with beets, mustard, rye bread and butter.

Potage Purée de Pois Cassés   *France*

Perfect winter food – typically French and comforting – which is even more delicious if you stir in some extra butter before serving and dribble some light cream on the top.

1½ cups (375ml) split green peas
3 cups (750ml) chicken stock
Bouquet garni of 1 bay leaf, 1 sprig thyme, 2
    sprigs parsley
1 ham bone
¼ pound (100g) blanched bacon
2 tablespoons butter
1 carrot, peeled and chopped
1 small onion, chopped
1 leek, finely chopped
3 lettuce leaves, coarsely chopped
Salt and pepper
½ pound (225g) ham, cut in coarse chunks

Cook the peas in the chicken stock with the *bouquet garni* and ham bone for 45 minutes.

Slowly cook the bacon until all the fat has run out. Remove the crisp bacon pieces. Add the butter to the bacon fat and lightly cook the vegetables until they begin to soften. Add to the peas and simmer gently for a further 20 minutes.

Remove the ham bone and herbs. Purée or sieve the soup and return to the pan. Reheat and season with salt and pepper. Add the ham cubes and serve hot.

Potage Soissonnaise   *France*

As the ingredients and method are so similar to the preceding recipe I have not given this in full.

Substitute soaked Great Northern beans for the split peas and increase the initial cooking time to 1 hour. Omit the thyme from the *bouquet garni*, and the ham cubes. The soup should be thick but smooth. Serve with a good lump of butter and some chopped parsley.

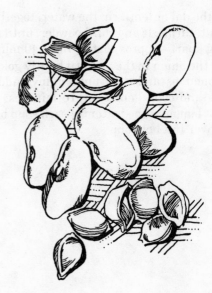

Lemon Dal Soup   *Anglo-Indian*

1½ cups (375ml) masur dal or red lentils
3 cups (750ml) water
1 teaspoon salt
1 teaspoon turmeric
½ teaspoon chilli powder
2 onions, chopped
1 tablespoon oil
4 tomatoes, chopped
1 lemon, thickly sliced
To serve:
Samosas (stuffed fried pastries), pakoras (vegetable fritters), or other Indian side dishes

Cook the dal or lentils in the water, together with the salt, turmeric and chilli powder, until soft and falling apart – approximately 30 to 40 minutes.

Fry the onion in the oil until lightly golden, add the tomatoes and cook until pulpy. Add to the lentils and cook gently for 10 minutes. Place a slice of lemon in the bottom of each soup bowl and pour over the hot soup.

Shurit Ads   *Egypt*

A smooth, buttery soup flavoured with cumin and lemon.

1 cup (250ml) red lentils, washed
1 onion, sliced
1 tomato, coarsely chopped
1 clove garlic, crushed
3 cups (750ml) chicken stock
1 small onion, finely chopped
3 tablespoons butter
Salt and pepper
1 teaspoon ground cumin
To serve:
1 lemon, cut into wedges

Cook the lentils with the sliced onion, tomato and garlic in the stock, simmering for 45 minutes.

Cook the chopped onion in 2 tablespoons butter until soft and beginning to brown.

Purée the soup, seasoning with salt, pepper and cumin. Stir the remaining butter into the soup and serve with the fried onions on top. Pass the lemon wedges and squeeze into the soup to taste.

Lentil and Watercress Soup   *France*

This soup is based on the French lentil and sorrel soup. Sorrel is particularly easy to grow. It has a lovely sharp, fresh flavour and is really better than the watercress substituted in this recipe.

¾ cup (150ml) green lentils
3 cups (750ml) chicken or vegetable stock
¼ pound (100g) watercress (or sorrel), chopped
2 tablespoons butter
Salt and pepper

Cook the lentils in the stock for 40 minutes.

Cook the watercress very gently in half the butter until soft. Sieve the lentils and the watercress and combine. Heat through and whisk in the remaining butter. Season and serve hot.

Lentil and Pasta Soup   *Italy*

Serve with warm bread for a winter lunch.

2 slices blanched bacon, diced
3 cloves garlic, crushed
1 small onion, chopped
3 tomatoes, peeled and chopped
1 stick celery, chopped
1 cup (250ml) brown lentils
4 cups (1 litre) water
Salt and pepper
¼ cup (50ml) broken spaghetti
Garnish:
Finely chopped parsley

Gently cook the bacon until the fat runs. Remove and reserve the bacon.

Very gently cook the garlic, onion, tomatoes and celery in the bacon fat until soft. Stir in the lentils, the water and a little salt and pepper. Bring to the boil and cook over a medium heat for 30 to 45 minutes or until the lentils are tender but not mushy.

Add the pasta and cook for a further 10 minutes. Serve very hot in deep bowls and sprinkle with parsley.

Turkish Lentil Soup   *Turkey*

A smooth, creamy soup especially popular with children.

4 tablespoons butter
1 onion, chopped
1 carrot, peeled and chopped
1 stick celery, chopped
1 cup (250ml) brown lentils
4 cups (1 litre) chicken stock
2 tablespoons flour
1 cup (250ml) milk
2 egg yolks, beaten together
Salt and pepper
1 teaspoon ground coriander
Juice of ½ lemon
To serve:
Toast or small garlic croûtons

Melt half the butter and cook the onion, carrot and celery gently until soft but not brown. Stir in the lentils and cover with the stock, bring to the boil and simmer gently for 40 to 45 minutes. Sieve or blend to a smooth purée.

Melt the remaining butter in a small pan and stir in the flour, but do not brown. Gradually stir in the milk and bring to the boil. Remove from the heat and gradually beat in the egg yolks.

Flavour the puréed lentils with salt, pepper, coriander and lemon juice and gradually beat in the white sauce. Reheat, but do not boil. Serve with toast or small garlic croûtons.

American Lentil Soup   *USA*

Substantial enough for a winter's lunch.

1 cup (250ml) brown lentils
3 cups (750ml) beef stock
1 cup (250ml) light beer or lager
3 slices bacon, chopped
2 cloves garlic, crushed
1 onion, chopped
1 green pepper, seeded and chopped
1 carrot, peeled and sliced
2 frankfurters, sliced
1 stick celery, chopped
¼ pound (100g) ham, diced
2 bay leaves
1 sprig thyme
Salt and pepper
Garnish:
Chopped parsley
Croûtons fried in bacon fat

Cook the lentils in the stock and beer for 30 minutes.

Fry the bacon over a low heat until the fat runs out. Add the garlic, onion and green pepper and fry gently until soft. Add to the lentils and cook for a further 15 minutes. Add all the remaining ingredients and continue simmering for 20 minutes. Season with salt and pepper to taste.

Serve sprinkled with parsley and, if you're not counting calories, croûtons fried in bacon fat.

Lentil Soup   *Spain*

1–2 tablespoons olive oil
1 large onion, chopped
2 cloves garlic, crushed
¾ pound (350g) tomatoes, peeled, seeded and chopped
1 red pepper, peeled, seeded and chopped
1 cup (250ml) brown lentils
Salt and pepper

Heat the oil in a large pan and cook the onion gently until soft but not brown. Add the tomatoes, red pepper, lentils and seasonings. Cover with water, bring to the boil, cover and simmer very gently for 1 hour.

Sieve or blend the soup, check the seasoning and serve hot.

Linsensuppe   *Germany*

Served with whole wheat or rye bread, this makes a warming winter lunch.

1 small ham hock or ¼ pound (100g) lean bacon pieces
1 cup (250ml) brown lentils
1 carrot, peeled and finely chopped
1 small celeriac (celery root), finely chopped
1 leek, finely chopped
3 cups (750ml) water
1 onion, finely chopped
Bacon fat or lard
1 tablespoon flour
2 frankfurters, sliced
Salt and pepper

Place the ham hock, lentils, carrot, celeriac, leeks and water in a pan, bring to the boil, cover and simmer gently for 30 minutes.

Fry the onions in the fat until only just gold. Stir in the flour and cook on a low heat for 1 minute. Add 2 tablespoons of the soup liquid, stirring well to blend the flour into a smooth paste. Gradually stir this into the soup and cook for a further 20 minutes or until the lentils are soft.

Remove the ham hock, dice the meat and return to the soup. If using bacon pieces they can be merely chopped into small dice. Add the sliced frankfurters, salt and plenty of freshly ground black pepper. Reheat and serve.

Cream of Lentil Soup

A very elegant-looking soup which is made from humble ingredients, with the exception of the cream.

1 cup (250ml) brown or green lentils
2 cups (500ml) ham or chicken stock
Bouquet garni of parsley, celery leaves and lemon
    thyme
1 onion, finely chopped
1 tablespoon butter
1 tablespoon flour
1 cup (250ml) milk
Salt and white pepper
Grated nutmeg
To serve:
4 tablespoons cream

Cook the lentils in the ham stock, together with the *bouquet garni*, for approximately 30 minutes. Purée.

Cook the onion in the butter until soft. Stir in the flour and, stirring continuously, add the milk. When well blended, add the puréed lentils.

Season with salt, white pepper and a little nutmeg and serve with a spoonful of cream on each serving.

Westphalian Lentil Soup *Germany*

1½ cups (375ml) brown lentils
1 onion, chopped
1 carrot, chopped
1 stick celery, chopped
1 leek, chopped
3 slices bacon, chopped
3 cups (750ml) chicken or ham stock
¼ pound (100g) German smoked slicing sausage,
    preferably Katenrauchwurst or Mettwurst
½ cup (125ml) light cream
3 tablespoons finely chopped parsley
Salt and pepper

Cook the lentils, onion, carrot, celery, leek, bacon and stock in a large pan for 1 hour, until the lentils and vegetables are soft. Add the sausage and continue cooking for 10 minutes.

Remove the sausage and cut into slices. Return to the soup. Stir in the cream and parsley and season to taste. Serve very hot with fresh bread.

The soup can be puréed before the sausage slices, cream and parsley are added, if you prefer.

Sopa de Garbanzos _Spain_

A simple soup with the unexpected addition of mint.

1 cup (250ml) chick-peas, soaked
3 cloves garlic
3 tablespoons chopped parsley
8 sprigs mint
Salt
4 tablespoons olive oil
3 cups (750ml) chicken stock
To serve:
Croûtons fried in oil

Cook the chick-peas for 1 hour in enough water to cover them. Meanwhile, using a mortar and pestle, crush the garlic, parsley and half the mint to a smooth paste. Add the salt and beat in the oil in a slow stream until it is thoroughly absorbed.

Drain the chick-peas and discard the liquid. Add them to the chicken stock and bring to the boil.

Place the herb mixture in a tureen or divide between individual soup bowls. Pour over the soup, stirring well. Tear the remaining mint leaves into large pieces and scatter on the top of the soup. Serve the croûtons separately.

Garbanzo Soup _Spain_

Another meal in itself. The ham bone is essential.

¾ cup (375ml) chick-peas, soaked
1 ham bone
¼ pound (100g) pork butt
1 level teaspoon paprika
1 Spanish onion, chopped
2 potatoes, peeled and sliced
¼ pound (100g) chorizo, or any small smoked sausages
2 tablespoons oil
Salt and pepper

Put all the ingredients in a large saucepan, cover with water and bring to the boil. Cover and simmer for 1 to 1½ hours.

Remove the ham bone and discard. Slice the sausage and pork and return to the soup.

Circassian Soup   *Turkey*

½ cup (125ml) chick-peas, soaked
1 carrot, peeled and chopped
1 stick celery, chopped
2 cups (500ml) chicken stock
2 tablespoons pearl barley
1 tablespoon butter
1 bay leaf
Salt, pepper and a pinch of cayenne pepper
1 onion, chopped
3 tomatoes, peeled and chopped
2 potatoes, peeled, boiled and mashed
1 tablespoon tomato paste
To serve:
Plain yoghurt

Cook the chick-peas, carrot, and celery in the stock for 15 minutes. Add the pearl barley, bay leaf, salt, pepper and cayenne and simmer for a further 30 minutes.

Fry the onion in the butter until lightly golden, add the tomatoes and cook for 1 minute.

Stir the potato into the soup and add the tomato paste, onion and tomato. Cook for 25 minutes. Serve hot with a bowl of yoghurt.

Zuppa de Ceci   *Italy*

Pigs' feet are still cheap, and make a good rich stock which complements the earthy taste of the chick-peas perfectly.

1 cup (250ml) chick-peas, soaked
1 onion, chopped
1 stick celery, chopped
4 tablespoons butter
2 pigs' feet, well washed
2 tomatoes, chopped
2 carrots, peeled and chopped
2 cloves garlic, crushed
1 bay leaf
½ cup (125ml) dry white wine
Salt and pepper
2 slices blanched bacon, diced
1 tablespoon tomato paste
Garnish:
Finely chopped parsley

Cover the chick-peas with cold water and cook for 40 minutes. Drain and reserve the liquid.

Lightly fry the onion and celery in the butter and when soft add the pigs' feet, turning them to coat with the butter. Add the tomatoes, carrots, garlic, bay leaf, wine, salt and pepper. Cover with the cooking liquid from the chick-peas, reserving the chick-peas until later. Cover and simmer for 45 minutes.

Gently cook the bacon until all the fat has run out, then stir in the tomato paste and a few spoonfuls of the soup. Stir well and add to the soup with the chick-peas.

Remove the pigs' feet and cut the meat into small pieces. Return to the soup, bring to the boil and serve sprinkled with parsley.

The traditional lamb soup that breaks the Muslim fast at sunset during Ramadan. Recipes vary from family to family, but the use of yeast as a thickening agent is peculiar to Morocco. However, this dish can also be left unthickened.

To serve this as a main meal rather than a filling soup, double the quantities.

1 tablespoon olive oil
½ pound (225g) lamb, cubed
½ teaspoon saffron
½ teaspoon ground cumin
¼ teaspoon turmeric
½ teaspoon ground ginger
½ teaspoon ground coriander
2 onions, chopped
1 carrot, peeled and chopped
2 tomatoes, peeled and chopped
3 cloves garlic, crushed
2 cups (500ml) chicken stock
A few lamb bones, if available
Salt and pepper
¼ teaspoon paprika
½ cup (125ml) chick-peas, soaked
Handful of parsley, chopped
¼ cup (50ml) long-grain rice
½ teaspoon dried yeast

Heat the oil and lightly brown the lamb, turning frequently until evenly coloured. Stir in the spices, onion, carrot, tomatoes and garlic. Cover with the chicken stock, add the bones, season with salt, pepper and paprika and bring to the boil. Add the chick-peas, cover and simmer for 1 hour 40 minutes.

Add the handful of chopped parsley and the rice and cook for a further 20 minutes. Discard the bones.

Remove a ladleful of the stock and leave in a small bowl until lukewarm. Stir in the yeast and leave for 5 minutes. Whisk into the soup and reheat, but do not boil. Serve in deep bowls, very hot.

La Soupe Normande
France

1 onion, sliced
2 tablespoons butter
6 sticks celery, diced
3 carrots, peeled and diced
3 cups (750ml) water
1 cup (250ml) small white beans or brown beans, soaked
2 tablespoons pearl barley
3 cloves
Salt and pepper
Garnish:
1 tablespoon parsley, chopped

Cook the onion in butter until lightly browned, then add the celery and carrots. Cover with the water and add the beans, barley and cloves. Bring to the boil, then simmer for 2 hours.

Sieve the soup and reheat. Season and serve sprinkled with finely chopped parsley.

Bean Sprout Soup

1 onion, sliced
1 carrot, peeled and chopped
2 sticks celery, chopped
1 tablespoon oil
¾ pound (350g) mixed fresh mung bean and aduki sprouts in equal quantities (about 6 cups or 1½ litres)
2 tomatoes, peeled and chopped
3 cups (750ml) vegetable or chicken stock
Salt and pepper

Cook the onion, carrot and celery in the oil until soft. Add most of the bean sprouts, reserving a few for garnish. Add the tomatoes and stock and season. Bring to the boil, cover and simmer for 30 minutes.

Sieve or purée in a blender. Check seasoning and serve sprinkled with remaining bean sprouts.

More of a stew than a soup, this can be served as a meal on its own. It should include chorizos (smoked, coarsely-chopped pork sausages flavoured with paprika, garlic and herbs), which are available at most delicatessens, but they can be replaced by any smoked garlic sausage although not the slicing variety.

The name 'fabada' comes from the word 'faba', the local name for the dried white broad beans usually sold as ful nabed. Served with corn bread, this is the favourite dish of north western Spain.

4 cups (1 litre) ful nabed beans, soaked in 4 cups (1 litre) water
2 cloves garlic, finely chopped
2 onions, chopped
1 pig's foot, well washed (optional)
¼ pound (100g) pork butt
6 ounces (175g) lean ham
1 teaspoon saffron threads, infused in 2 tablespoons hot water
6 ounces (175g) chorizo or smoked garlic sausage
¼ pound (100g) blood sausage, sliced
Salt and pepper

Combine the bean soaking water with an equal amount of fresh water. Put the beans, garlic and onions in a large pot, pour over the water and bring to the boil. Add the pig's foot, pork butt and ham. Cover and simmer for 2 hours.

Boil the chorizos for 5 minutes, drain and slice. Stir the saffron, chorizos, blood sausage and seasoning into the soup and cook gently for 20 minutes.

Discard the pig's foot. Remove the bacon and ham, cube coarsely and return to the soup. Reheat and serve very hot in deep bowls.

Ful Nabed Soup   *Egypt*

A very simple soup but one which is enhanced by the distinctive flavour of ful nabed.

1 cup (250ml) ful nabed, soaked overnight
3 cups (750ml) water
1 tablespoon olive oil
Salt and pepper
Garnish:
Chopped parsley
Juice of ½ lemon

Cook the beans in the water for 1½ to 2 hours. Purée or sieve and return to the pan. Bring slowly to the boil, stirring in the oil, salt and pepper (you can use white pepper if you object to black flecks in a white soup).

Serve hot garnished with chopped parsley and with a good squeeze of lemon juice. Alternatively, hand round quartered lemons separately.

Fasalada Sopa   *Spain*

1 onion, sliced
1 clove garlic, crushed
3 tablespoons olive oil
1 tablespoon tomato paste
½ teaspoon dried thyme
1 cup (250ml) fava beans (ful medames)
Juice of ½ lemon
Handful of chopped parsley

Cook the onion and garlic gently in the oil until soft but not brown. Stir in the tomato paste and thyme. Add the beans and cook for 3 hours in enough water to cover by 1 inch (2½cm).

Sieve coarsely or purée in a blender at slow speed. Stir in the lemon juice, salt and parsley. Serve hot.

Caldo Gallego   *Spain*

There are many variations of this dish – sometimes the broth is served separately and the vegetables and meats served with potatoes. In Pote Gallego chicken is added and, in other versions, pork.

The inclusion of turnip tops is purely Galician and may not be appreciated by everyone. They have the same slightly bitter tang as kale (which can be used instead).

½ cup (125ml) Great Northern beans, soaked
1 small onion, chopped
2 slices blanched bacon
6 ounces (175g) slab bacon, cubed
¼ pound (100g) chorizo or similar smoked garlic
   sausage
1 potato, peeled and diced
6 ounces (175g) turnip tops, shredded
Salt and pepper

Put the beans in a heavy saucepan with the onions, bacon and cubed bacon. Cover with water by 2 inches (5cm) and bring to the boil. Cover and simmer for 1½ hours.

Add the sausage and potatoes and cook for 10 minutes. Add the turnip tops to the soup, season and cook for 20 minutes.

Remove the sliced bacon and the sausage. Slice the sausage, return to the soup and serve.

Bean Soup   *USA*

Home-made bread, cheese and fruit make this a complete meal.

1 cup (250ml) Great Northern beans, smoked
2 cups (500ml) ham or chicken stock
1 onion, sliced
1 carrot, peeled and chopped
2 sticks celery, chopped
1 tablespoon oil
1 teaspoon dry mustard
1 teaspoon brown sugar
1 teaspoon marjoram
1 tablespoon tomato paste
¼ pound (100g) ham, diced
2 frankfurters, sliced
Salt and pepper

Cook the beans in the stock for 45 minutes to 1 hour.

Cook the vegetables in the oil until soft but not brown. Stir the mustard, sugar and marjoram into the vegetables.

Purée the beans and return to the stock, adding the vegetable mixture, tomato paste, ham and frankfurters. Simmer gently for 15 minutes. Season with pepper and a little salt, depending on the saltiness of the original stock.

Purée de Judias Blancas *Spain*

A smooth soup for garlic lovers – but it must be *olive* oil.

1 cup (250ml) Great Northern beans, soaked
4 cloves garlic, crushed
1 small onion, sliced
1 tablespoon olive oil
4 slices blanched bacon, diced
1 tablespoon butter
½ teaspoon paprika
Salt

Cook beans in enough water to cover them by 1 inch (2½cm), adding the garlic once the liquid has come to the boil. Simmer for 1 hour.

Cook the onion in the oil until soft but not brown. Add to the beans and purée with the bean liquid. Return to the pan and keep warm.

Lightly fry the bacon in the butter, remove from heat and stir in the paprika. Stir into the soup, making sure the paprika is well distributed. Season with salt to taste. Serve very hot.

Bean and Pasta Soup *Italy*

1 cup (250ml) Great Northern beans, soaked
4 cups (1 litre) water
½ pound (225g) bacon, finely diced
1 onion, finely chopped
2 sticks celery, finely chopped
2 cloves garlic, crushed
1 tablespoon oil
Salt and pepper
¼ pound (100g) pork butt
½ cup (125ml) broken pasta (spaghetti, vermicelli, tagliatelle, etc.
To serve:
Freshly grated Parmesan cheese

Cook the beans in the water for 30 minutes. Drain and reserve the liquid.

Fry the bacon, onion, celery and garlic in the oil until the vegetables are soft. Add the beans to the pan, stirring, and cook over a gentle heat for 2 minutes. Add the bean cooking liquid and salt and pepper and bring to the boil. Add the pork butt, cover the pan and simmer for 1 hour.

Remove the pork. Crush the beans coarsely in the soup.

Add the pasta and cook for 10 minutes until tender. Serve very hot with freshly grated Parmesan cheese.

French Chick-pea Soup *France*

A simple soup, easily made and always popular.

1 tablespoon olive oil
1 onion, chopped
1 large or 2 small leeks, chopped
2 tomatoes, chopped
1 cup (250ml) chick-peas, soaked
3 cups (750ml) water or stock
To serve:
Croûtons or hot garlic bread

Heat the oil and gently cook the onion, and tomatoes until very soft. Add the chick-peas and stock. Bring to the boil and simmer for 1 hour or until the chick-peas are soft.

Sieve the soup, which should be fairly thick, then return to the pan and reheat. Serve very hot.

Sopa de Grão *Portugal*

1 cup (250ml) chick-peas, soaked
1 onion, chopped
3 slices blanched bacon, chopped
½ pound (225g) spinach, well washed
2 tablespoons olive oil
2 ounces (50g) chorizo or garlic sausage
Salt and pepper

Put the chick-peas, onion, bacon, oil and sausage in a saucepan, cover with water and add a pinch of salt. Bring to the boil, cover and simmer for 1 hour.

Remove the sausage and cut into slices or dice. Purée the soup. Return to the saucepan and check the seasoning.

Add the sausage and the finely chopped spinach. Bring to the boil and cook for 5 minutes. Serve in deep bowls.

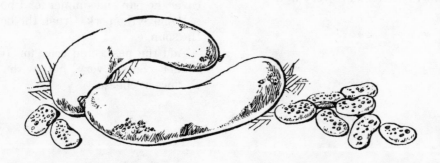

Córba (1)   *Bulgaria*

Córba (or Ciorba) are the thick Balkan stews of Rumania and Bulgaria. They are often served acidulated with either vinegar, lemon or sauerkraut juice. Serve as a main meal with a green salad.

1 cup (250ml) Great Northern beans, soaked
4 cups (1 litre) water
2 onions, chopped
1 clove garlic, crushed
½ pound (225g) bacon, chopped
3 potatoes, peeled and cubed
3 carrots, peeled and chopped
¼ pound (100g) smoked sausage, sliced
Juice of 1 lemon

Cook the beans in half the water for 1 hour. Sieve and add the remaining water, onions and garlic. Cook for a further 30 minutes, uncovered, over a moderate heat.

Cook the bacon slowly until the fat has run out. Remove the bacon and fry the vegetables in the bacon fat.

Add the bacon and vegetables to the soup. Bring to the boil, add the sausage and heat through. Add a dash of lemon juice to taste.

Córba (2)   *Bulgaria*

1 cup (250ml) Great Northern beans, soaked
1 carrot, peeled and chopped
1 small celeriac (celery root), chopped
1 tablespoon tomato paste
2 tablespoons olive oil
1 small dried chilli, whole but seeded
¼ pound (100g) bacon, chopped
Salt and pepper
Garnish:
Finely chopped mint and parsley

Cover the beans with cold water and add the vegetables, tomato paste, oil, chilli and bacon. Simmer for 1 hour.

Remove the chilli. Season and serve hot with the chopped fresh herbs sprinkled over.

L'Ouillade — *Catalan*

1½ cups (375ml) Great Northern beans, soaked
2 carrots, peeled and chopped
4 potatoes, peeled and chopped
1 small cabbage, coarsely shredded
2 onions, finely chopped
1 clove garlic, crushed
¼ pound (100g) slab bacon
1 ham bone
Salt and pepper
1 sprig thyme
1 tablespoon finely chopped parsley
1 clove garlic, chopped
Lard or bacon fat

Cook the beans for 1 hour. Drain.

Place the carrots, potatoes, cabbage, 1 onion, crushed garlic, bacon, ham bone, salt, pepper and thyme in a pan with enough water to just cover and boil for about thirty minutes or until the vegetables are just tender.

Remove the bone and bacon. The bacon can be eaten at another meal or chopped and returned to the soup.

Cook the parsley, the chopped garlic and the remaining onion lightly in the bacon fat or lard.

Add the beans to the vegetables with enough of their cooking liquid to make a fairly thick soup. Stir in the fried onion and herb mixture and serve very hot.

Beans and Sauerkraut Soup — *Bulgaria*

An unusual soup for sauerkraut buffs.

1 cup (250ml) Great Northern beans, soaked
4 cups (1 litre) chicken stock
2 potatoes, peeled and cubed
½ pound (225g) bacon, cubed
½ pound (225g) sauerkraut (tinned or fresh)
2 cloves garlic, crushed
2 tablespoons oil
2 tablespoons flour
Salt and pepper
To serve:
Plain yoghurt

Cook the beans in the stock for 1 hour, adding the potatoes after 30 minutes.

Simmer the bacon and sauerkraut in a covered pan for 30 minutes. Purée the beans and potatoes, returning them to the cooking liquid.

Lightly cook the garlic in the oil until soft but not coloured. Stir in the flour and cook until it begins to brown. Add 3 tablespoons of the bean liquid, stirring quickly to a smooth paste. Stir into the soup and bring to the boil, still stirring.

Drain the sauerkraut and bacon and add to the soup. Heat through and serve with a large dollop of yoghurt.

Zuppa di Fagioli alla Toscana *Italy*

Classically simple – but it needs good olive oil.

1 cup (250ml) Great Northern beans, soaked
4 cups (1 litre) water
3 cloves garlic, chopped
1 tablespoon olive oil
2 tablespoons finely chopped parsley
Salt and pepper

Cook the beans in the water for 1 hour. Drain and reserve the liquid. Purée half the beans.

Cook the garlic lightly in the oil. Remove from the heat and stir in the parsley.

Combine the beans, bean purée and cooking liquid in a pan, heat through and stir in the garlic/parsley mixture. Season and serve hot.

Bohnensuppe *Germany*

¾ cup (175ml) Great Northern beans, soaked
3 cups (750ml) beef, chicken or ham stock
1 carrot, peeled and chopped
1 stick celery, chopped
1 small ham hock (optional)
1 large onion, chopped
1 tablespoon butter
1 tablespoon flour
Salt and pepper

Cook the beans in the stock with the carrot, celery and ham hock for 1 hour. Drain and reserve the liquid.

Remove the ham hock and use for another dish. Purée half the beans and half the vegetables. Stir back into the whole beans and vegetables and return to the pan together with the liquid.

Fry the onion in the butter until golden then stir in the flour. Cook gently for 1 minute then stir in 3 tablespoons of the soup. Stir into the soup, reheat and adjust the seasoning.

Swiss Bean Soup — *Switzerland*

A thick main meal soup.

1 onion
4 cloves
¾ cup (175ml) Great Northern beans, soaked
¾ cup (175ml) pearl barley, soaked overnight
½ pound (225g) ham or bacon, in one piece
½ pound (225g) lean chuck or stewing beef
1 small bay leaf
Salt and pepper
Ham stock or water

Spike the onion with the cloves. Place all the ingredients in a large saucepan, adding enough stock or water to cover by approximately 3 inches (7½cm). Bring to the boil, cover and simmer for 2½ to 3 hours.

Remove the onion. Remove and slice the ham and beef and return to the soup. Check seasoning and serve very hot.

Paprika Bean Soup — *Bulgaria*

1 cup (250ml) Great Northern beans, soaked
2 onions, sliced
2 cloves garlic, crushed
1 tablespoon flour
2 tablespoons olive oil
1–2 teaspoons paprika
To serve:
4 slices toasted bread

Cook the beans for 1 hour in their soaking water. Drain and purée them. Thin the purée with 2 cups (500ml) water. Add the onion and garlic and cook gently for 1 hour.

In a small bowl whisk together the flour, oil and paprika. Whisk this mixture into the soup and bring to the boil. Put one slice of toast in each soup bowl and pour the soup over.

Soupe au Pistou

Related to the Italian Minestrone Genovese, this Provencal soup is very filling and should be served as a main course with plenty of hot French bread. To be authentic, fresh basil should be used (it grows easily on a windowsill), but the dried variety will also give a pleasant-tasting soup.

½ cup (125ml) Great Northern beans, soaked
4 cups (1 litre) water
1 small onion, chopped
1 tablespoon olive oil
3 tomatoes, peeled, seeded and chopped
3 carrots, peeled and diced
1 leek, chopped
1 stick celery, including leaves, chopped
2 potatoes, peeled and diced
Salt and pepper
2 zucchini, diced
¼ pound (100g) green beans, sliced
1 pinch saffron threads
¼ cup (50ml) broken spaghetti
Pistou:
3 cloves garlic
2 tablespoons chopped fresh basil, or 2 teaspoons dried
3 teaspoons tomato paste
½ cup (125ml) freshly grated Parmesan cheese
2 tablespoons olive oil

Cook the beans in the water for 1 hour. Drain and reserve the cooking liquid.

Cook the onion in the oil until soft and light gold, add the tomatoes and cook for 2 minutes, stirring. Add half the bean liquid, the carrots, leeks, celery, potatoes, salt and pepper. Bring to the boil and simmer for 15 minutes.

Add the zucchini, green beans and the remaining liquid. Continue cooking for 5 minutes. Bring to the boil and add the saffron and spaghetti. Lower the heat and simmer for 15 minutes.

Meanwhile, prepare the Pistou by crushing the garlic and basil together in a mortar to a pungent, green paste. Blend in the tomato paste and the cheese. Add the oil slowly, as for mayonnaise, until you have a thick paste. Thin with 2 tablespoons of soup liquid. Although the pistou is then normally put into the tureen and the soup stirred in, I usually serve it in a separate bowl. It can then be added at the table to suit individual tastes.

Serve with a bowl of freshly grated Parmesan cheese.

This winter soup/stew from the Pyrenees should really contain *confit d'oie* (preserved goose), but this version relies on slab bacon and smoked garlic sausage to give it flavour and body. The bacon and sausage are then sliced and served separately with French mustard, either to follow the soup or cold the next day. The addition of chestnuts gives this recipe a superb flavour (for which I am indebted to my brother who lived at one time in the Pyrenees), as does the Hâchis, which is borrowed unashamedly from Elizabeth David's *French Country Cooking*.

Be warned – this is not a meal for slimmers or those with small appetites!

½ cup (125ml) Great Northern beans, soaked
½ pound (225g) slab bacon
¼ pound (100g) garlic sausage, preferably the small type, in one piece
1 ham bone, if available
2 potatoes, peeled and cubed
2 leeks, sliced
1 onion, sliced
2 carrots, peeled and cubed
2 turnips, peeled and cubed
1 red pepper, seeded and cut into rings
⅓ cup (75ml) peas, fresh or frozen
2 cloves garlic, crushed
1 pound (450g) crisp cabbage, shredded
Bouquet garni of 1 bay leaf, 3 sprigs parsley, 1 sprig thyme and 1 sprig marjoram
¼ pound (100g) chestnuts, roasted or boiled, if available

Hâchis:
1 slice bacon, chopped
1 clove garlic, chopped
1 small onion, finely chopped
Few sprigs parsley
To serve:
Chunks of coarse brown bread

Cook the beans for 1 hour.

Meanwhile, cover the bacon piece and the ham bone with 4 cups (1 litre) water and bring to the boil. Add the vegetables, cabbage and *bouquet garni*, cover and simmer for 45 minutes. Add the sausage and chestnuts and continue simmering for 20 minutes.

Meanwhile, prepare the *hâchis* by slowly cooking the bacon until the fat begins to run. Add the garlic, onion and parsley and cook lightly in the fat.

Discard the *bouquet garni* and ham bone from the soup. Remove the bacon and sausage and eat separately (see Introduction). Stir the *hâchis* into the soup.

Drain the beans and add to the soup, thinning it with a little of the bean liquid if necessary. However, the soup should really be thick enough for a spoon to stand up in it.

It seems rather superfluous to give a recipe for Minestrone when so many exist. However, they vary in their authenticity (and taste), and I can vouch for both in this version.

½ cup (125ml) Great Northern beans, soaked
3 cups (750ml) chicken stock or water
2 slices blanched bacon, finely diced
1 onion, finely chopped
1 leek, finely chopped
2 tomatoes, chopped
1 potato, peeled and diced
2 carrots, peeled and diced
2 zucchini, diced
2 sticks celery, diced
½ cup (125ml) fresh or frozen peas
1 bay leaf, 2 sprigs parsley and 1 sprig thyme, tied
    together
Salt and pepper
2 tablespoons long-grain rice
Garnish:
Handful of parsley, finely chopped
1 clove garlic, finely chopped
Freshly grated Parmesan cheese

Cook the beans in half the water or stock for 40 minutes, or until just tender.

In a small frying pan cook the bacon slowly so the fat runs out. When the bacon pieces are crisp, remove them and drain on paper towels. Cook the onion and leek gently in the bacon fat until soft, but not browned. Add the tomatoes and cook for 1 minute.

In a separate pan melt the butter and lightly cook the potatoes, carrots, zucchini, celery and peas, if fresh. (If frozen peas are used, add them 5 minutes before the end.)

Combine the two separate groups of vegetables, adding the second half of the stock or water. Bring to the boil and add the herbs. Season and simmer, with the lid askew, for 30 minutes.

Add the rice and drained beans and cook 15 minutes longer. Remove the herbs and check the seasoning.

Mix together the parsley and garlic. Serve the soup sprinkled with this herb mixture, the crisp bacon pieces and plenty of freshly grated Parmesan.

Black Bean Salad

1 cup (250ml) black beans, soaked
1 tablespoon cider vinegar or lemon juice
2 tablespoons oil
Salt and pepper
1 onion, chopped
2 large tomatoes, chopped
2 cloves garlic, crushed

Cook the beans for 45 minutes or until tender. Drain well.

Stir in the vinegar, oil, salt and pepper while the beans are warm.

Cool and add the onion, tomatoes and garlic. Serve chilled.

Lobia Dal Salad *India*

1 cup (250ml) black eyed peas, soaked
2 tablespoons oil
1½ tablespoons wine or cider vinegar
2 cloves garlic, crushed
½ teaspoon paprika
Salt and pepper

Cook the peas in unsalted water to cover for 45 minutes, keeping them whole.

Drain well and stir in the remaining ingredients while the peas are still warm.

African Bean Salad *Africa*

1 cup (250ml) black eyed peas, soaked
1 onion, chopped
1 clove garlic, chopped
1 tablespoon oil
2 tablespoons tomato paste
6 ounces (150g) cooked shrimp, peeled and coarsely
  chopped
¼ pound (100g) ham, diced
Salt
1 tablespoon lemon juice

Cook the peas for 45 minutes. Drain well.

Cook the onion and garlic in the oil until soft but not coloured. Stir in the tomato paste, shrimp and ham. Remove from the heat and stir into the peas.

Season to taste, sprinkle with lemon juice and serve cold.

French Chick-pea Salad *France*

Simplicity itself.

1 cup (250ml) chick-peas, soaked
3 tablespoons olive oil
1 tablespoon lemon juice
2 shallots, finely sliced
Salt and pepper

Cook the chick-peas until tender, about 45 minutes. Drain well.

Mix together the oil, lemon, salt and pepper. Stir into the warm chick-peas. Cool to room temperature and add the shallots.

Salatet Hummus *Arab*

1 cup (250ml) chick-peas, soaked
1 small onion, chopped
1 clove garlic, crushed
2 tablespoons olive oil
1 tablespoon parsley, chopped
Juice of ½ lemon
Salt and cayenne pepper

Cook the chick-peas for 45 minutes and drain well.

Mix together the onion, garlic, oil, parsley, lemon juice, a pinch of salt and a pinch of cayenne. Toss the warm chick-peas in the dressing and cool to room temperature.

Chick-pea and Bulgur Salad *Lebanon*

This is a variation on the Lebanese Tabbouleh, a marvellously refreshing salad and a personal favourite.

½ cup (125ml) chick-peas, soaked
½ cup (125ml) bulgur (cracked wheat), soaked in cold water for about 45 minutes.
1 onion, finely chopped
3 tablespoons olive oil
Juice of 1 lemon
Salt and pepper
Handful of parsley, finely chopped
To serve:
Crisp lettuce
1 tomato, thinly sliced
¼ cucumber, thinly sliced

Cook the chick-peas for 1 hour until soft. Drain well and skin if you have time.

Squeeze the bulgur dry with your hands, or wring out in a piece of muslin or teatowel. Spread out to dry on a cloth or paper towel.

Mix the onion, oil, lemon, salt and pepper, crushing the onion slightly. Mix in the bulgur, squeezing and slightly kneading with your hands so that the flavours really penetrate. Stir in the chick-peas and parsley. Serve on lettuce leaves and decorate with tomato and cucumber.

Soya and Mushroom Salad

The permutations on this are endless, as any combination of good-textured vegetables produces a filling and nourishing salad.

1½ cups (375ml) soya beans, soaked
1 tablespoon oil
1 tablespoon lemon juice
1 clove garlic, crushed
Salt and pepper
3 sticks celery, chopped
6 ounces (175g) button mushrooms, sliced
4 green onions, finely chopped

Cook the beans for 2 hours until soft. Drain well.

Mix together the oil, lemon juice and garlic. Season well. Toss the beans and vegetables in the dressing and serve.

Ensalada de Habas   *Spain*

Although the recipe is for lima beans, haricot beans or soya beans may be used.

1 cup (250ml) lima beans, soaked
1 onion, finely sliced
1¼ cups (300ml) green peas, cooked
2 tablespoons olive oil
2 tablespoons wine vinegar
Salt and pepper

Cook the beans in unsalted water until just tender, 45 minutes to 1 hour. Drain well and add the onion and peas.

Mix the oil, vinegar, salt and pepper and stir into the beans while still warm. Serve chilled.

This can be extended with the addition of some crisp lettuce leaves, sliced tomatoes and diced cucumber.

Lentil and Anchovy Salad   *France*

1½ cups (375ml) brown lentils
1 large onion
1–2 tablespoons olive oil
Salt and pepper
Juice of ½ lemon
1 clove garlic, crushed
Garnish:
1 2½-ounce (62g) can anchovy fillets
Finely chopped parsley

Cook the lentils in unsalted water for 20 to 30 minutes or until tender. Drain. Slice about ⅔ of the onion into fine rings and chop the rest finely.

Heat the oil and gently cook the chopped onion and the crushed garlic until soft. Add the lentils to the pan and season with salt, pepper and lemon juice, being careful with the salt because of the anchovies. Add the crushed garlic and stir for a few minutes.

Transfer to a serving dish and cool. Decorate with the onion rings, anchovies and parsley. Serve either at room temperature or well chilled.

Lentil and Yoghurt Salad   *India*

This is based on an Indian dish which is usually served hot. Ideally you should use thick, home-made yoghurt but the commercial variety can be used as well. Serve this with spiced vegetables and rice.

1½ cups (375ml) masur dal, soaked for 1 hour
1 heaped teaspoon ground cumin
½ teaspoon ground cinnamon
1 teaspoon paprika
1 teaspoon salt
Seeds of 1 cardamom pod, crushed
¾ cup (175ml) plain yogurt

Cook the dal in lightly salted water for 30 minutes, then drain.

Stir the spices and salt into the yoghurt. Mix the lentils gently with the yoghurt and leave until cold.

Salade de Lentilles _France_

This is a decorative salad from Provence and needs little other than bread and fresh fruit to make a quick summer meal.

1¼ cups (300ml) brown or green lentils
2 tablespoons olive oil
1 tablespoon wine vinegar
Salt and pepper
2 shallots, finely chopped
1 tablespoon strong French mustard
3 tomatoes, chopped
1 small red pepper, seeded and sliced into rings
¼ cucumber, diced
1 tablespoon finely chopped parsley
Garnish:
3 hard-cooked eggs, sliced

Cook the lentils in unsalted water for 25 minutes until tender but not mushy. It is preferable to shorten the cooking time if using a pressure cooker to avoid overcooking, unless you have used the lentils in question before and can judge their cooking time.

Drain and, while still hot, mix with the oil and vinegar and add salt and pepper to taste. Stir the shallots and mustard into the lentils, making sure they are evenly distributed. Leave to cool.

Stir the vegetables and parsley gently into the salad. Decorate with sliced eggs and serve.

Lentil and Rice Salad

This is a good winter meal, especially if teamed with a tomato and basil salad. A good way of using up left-over risotto (if you have a family that leaves anything!) is to add cooked lentils and a dressing of oil and lemon.

1 onion, chopped
1 tablespoon olive oil
½ cup (125ml) long-grain rice
1 cup (250ml) lentils
1½ cups (375ml) chicken stock
Juice of 1 lemon
1 tablespoon chopped parsley
Salt and pepper
Garnish:
3 hard-cooked eggs, sliced

Cook the onion in the oil for 10 minutes or until soft but not coloured.

Stir in the rice and lentils, turning them in the oil until lightly coated. Pour in the chicken stock, bring to the boil, cover and cook gently for 25 to 30 minutes. Cover the pan with a clean teatowel, return the lid and leave for 10 minutes off the heat. The rice must be dry and fluffy.

Leave until cold, then carefully stir in the lemon juice, parsley, salt and pepper. Chill and serve with the eggs arranged around the top of the dish.

Lentil Salad   *France*

A good, simple lunch.

1 cup (250ml) brown or green lentils
Salt
2 tablespoons olive oil
1 tablespoon wine vinegar
2 teaspoons Dijon mustard
Pepper
2 tomatoes, peeled, seeded and coarsely chopped
2 shallots, finely chopped
2 tablespoons chopped parsley
Garnish:
2 hard-cooked eggs, sliced or quartered

Cook the lentils in highly salted water for 30 to 45 minutes. Drain well.

Blend together the oil, vinegar, mustard, salt and pepper and pour over the warm lentils. Allow to cool and stir in the tomatoes, shallots and parsley. Decorate with the eggs.

Lentil Salad   *North Africa*

Good on its own or with flat pitta bread, to follow an egg dish.

1 cup (250ml) brown lentils
6 tablespoons olive oil
Juice of 1 large lemon
2 cloves garlic, crushed
Salt and pepper
Garnish:
Finely chopped parsley

Cover the lentils with water and cook over a medium heat for 30 to 45 minutes, checking occasionally that the lentils are still whole.

Drain the lentils and while still warm pour over a dressing of oil, lemon juice, crushed garlic, salt and pepper. Cool and sprinkle with the parsley.

Fagioli Toscanelli con Tonno   *Italy*

The perfect combination of beans, oil and tuna – a justifiably famous dish from Tuscany. Although always listed as an antipasto, I serve it in large bowls (and large quantities) for a favourite summer or winter lunch. It must be olive oil, the best you can afford and, if you can get them, cannellini beans. Although not strictly traditional, I sometimes add quartered hard-cooked eggs and black olives to this.

1 cup (250ml) Great Northern or cannellini beans, soaked
3 tablespoons good olive oil
2 teaspoons lemon juice
Salt and pepper
1 small onion, finely sliced
Chopped parsley
1 7-ounce (198g) can tuna, preferably packed in olive oil

Cook the beans in unsalted water for 45 minutes. Drain. Mix with the oil and lemon juice while still warm. Season and leave to cool.

Stir in the onion and parsley and pour into a flat earthenware bowl.

Drain the tuna (reserving the olive oil), arrange chunks of tuna on top of the beans and pour the olive oil over the salad. If other packing liquid is used for the tuna, discard it.

La Salade à la Bretonne   *France*

A very simple salad from Brittany which goes well with cold meat.

1 cup (250ml) Great Northern beans, soaked
2 tablespoons oil
2 tablespoons wine vinegar
Salt and pepper
2 cloves garlic, crushed
1 small head endive

Cook the beans in unsalted water for 1 hour until tender. Drain.

Make a sharp vinaigrette dressing with the oil and vinegar, seasoning well. Toss the beans and garlic in the dressing. Cool to room temperature.

Wash the endive and dry with a clean cloth. Tear the leaves into large shreds (rather than cutting them) and toss with the other ingredients. Serve at once, while the endive is still crisp.

Fasul *Bulgaria*

1 cup (250ml) Great Northern beans, soaked
2 onions, chopped
2 tablespoons oil
1 tablespoon tomato paste
Chopped parsley
2 tablespoons water
Salt and pepper
To serve:
Slices of smoked sausage
Mixed pickles

Cook the beans in unsalted water for 1 hour. Drain.

Cook the onions in the oil until soft but not coloured, then stir in the tomato paste, parsley and water. Cook gently for 10 minutes. Stir into the beans and season to taste.

Chill and serve with the sliced sausage and pickles.

Haricot Salad with Herb Dressing *France*

Another simple French salad to serve as a side dish. Personally I prefer to use basil, as the liquorice taste of the fennel is not to everyone's liking.

1 cup (250ml) Great Northern beans, soaked
2 tablespoons oil
2 tablespoons vinegar
Salt and pepper
Handful of fresh basil or fennel leaves, finely chopped

Cook the beans for 1 hour until tender. Drain well and dress, while hot, with oil, vinegar, salt and pepper to taste and basil or fennel. Allow to cool to room temperature and serve.

Kuru Fasulya Salatasi   *Turkey*

A simple, colourful salad.

1 cup (250 ml) Great Northern beans, soaked
3 tablespoons olive oil
3 tablespoons wine vinegar
1 onion, finely sliced
4 tomatoes, quartered
1 red pepper, cut into rings or peeled, seeded and
   sliced
4 black olives, pitted and chopped
Small bunch of parsley, chopped
Salt and pepper

Cook the beans in unsalted water for 45 minutes. Drain well and combine with the oil, vinegar, onion, tomato, pepper, olives, parsley, salt and pepper. Toss lightly and chill.

La Salade de Haricot Blancs   *France*

This recipe is derived from one found in an old French recipe book from the turn of the century.

1 cup (250 ml) Great Northern beans, soaked
2 tablespoons olive oil
1 tablespoon wine vinegar
2 tablespoons chopped parsley
1 tablespoon chives, chopped
2 tablespoons chopped fresh basil (or any other
   fresh herb except sage or rosemary)

Cook the beans in unsalted water for 45 minutes or until tender. Drain and purée, adding a little of the cooking liquid. While still warm, beat in the other ingredients and allow to cool to room temperature.

Serve with any green salad – lettuce, chicory, endive or, to quote the original recipe, fennel.

Piaz   *Turkey*

This makes a good summer lunch.

1 cup (250ml) Great Northern beans, soaked
4 tablespoons olive oil
Juice of ½ lemon
Salt and pepper
Garnish:
1 tomato, sliced
2 hard-cooked eggs, quartered
8 black olives

Cook the beans in unsalted water for 45 minutes
or until tender but not broken. Drain well and
add the oil, lemon juice, salt and pepper. Cool.

Decorate with the tomato, the quartered eggs
and the olives.

Lobia Tkemali   *USSR*

This Caucasian dish, with its spicy damson dressing, makes a welcome change at Christmas to
help the inevitable cold poultry. Dried basil isn't
very satisfactory for this dish, but if you have to
use it, add it to the vinegar at the beginning.

1 cup (250ml) red kidney beans, soaked
3 tablespoons plum jam
1½ tablespoons vinegar
1 clove garlic
5 fresh basil leaves, finely chopped, or 2 teaspoons
    dried basil
¼ teaspoon cayenne pepper
Garnish:
Finely chopped parsley

Cook the beans in unsalted water for 45 minutes
to 1 hour. Drain well and cool.

Put the jam, vinegar and, if you are using it,
dried basil in a saucepan and bring to the boil,
stirring to dissolve the jam. Crush the garlic,
fresh basil (if available) and cayenne to a paste
and beat into the jam. Add salt to taste and pour
over the beans. Stir to distribute the sauce.

Leave for at least an hour, then serve chilled
with the chopped parsley sprinkled over.

Tahina Bean Salad

1¼ cups (300ml) red kidney or pinto beans, soaked
2 tablespoons soy sauce, preferably Tamari
1 tablespoon oil
2 tablespoons tahina paste
2 tablespoons lemon juice
Garnish:
Lemon slices

Cook the beans in unsalted water for 1 hour until tender. Drain.

Mix together the other ingredients and stir into the beans, adjusting the dressing to suit your taste. Cool to room temperature and serve garnished with lemon slices.

Mexican Bean Salad   *Mexico*

1 cup (250ml) red kidney beans, soaked
1 tablespoon olive oil
2 tablespoons wine vinegar
Salt
1 teaspoon chopped fresh basil
1 clove garlic, crushed
1 shallot (or ½ small onion), grated
½ teaspoon Tabasco sauce
2 tablespoons tomato paste

Cook the beans until tender, about 45 minutes. Drain. Mix all the other ingredients together and stir into the beans. Serve chilled.

Red Bean Salad

1 cup (250ml) red kidney beans, soaked
4 tablespoon olive oil
4 tablespoons cider vinegar
Salt and pepper
1 large onion, chopped
1 tomato, chopped
To serve:
Finely chopped parsley
1 red pepper, skinned, seeded and cut into strips

Cook the beans for 45 minutes to 1 hour or until soft. Drain well and mix in the oil and vinegar while still warm. Season well and add the chopped onion and tomato.

Sprinkle with the parsley and red pepper. Serve chilled.

Rajma Dal Salat   *India*

1 cup (250ml) red kidney beans, soaked
1 clove garlic, crushed
Juice of 1 lime
1 tablespoon oil
1 tablespoon chopped parsley
3 green onions, chopped
Salt

Cook the beans in unsalted water for 45 minutes. Drain well.

Mix the garlic with the lime juice, oil, parsley, onion and a little salt. Stir in the beans and leave until cool.

Three Bean Salad　　　　*India*

Soak and cook the beans separately if you wish to retain their individual colours.

⅓ cup (75ml) chick-peas, soaked
⅓ cup (75ml) black eyed peas, soaked
⅓ cup (75ml) red kidney beans, soaked
2 cloves garlic, crushed
2 tablespoons oil
1 small onion or shallot, finely chopped
½ teaspoon ground cumin
½ teaspoon ground coriander
3 tablespoons lemon juice
Salt and pepper

Cook the beans for 45 minutes or until just tender. Drain well.

Mix together the garlic, oil, onion, spices, lemon juice, salt and pepper and pour over the warm beans. Cool to room temperature, cover and chill for at least 1 hour.

A finely chopped fresh chilli can be added, but use with caution.

Mixed Bean Salad　　　　*USA*

For this recipe I prefer to cook the beans separately, as both the red beans and the chick-peas can colour the white beans. Obviously it is cheaper to cook them all together, but the overall colours will not be so distinct.

⅓ cup (75ml) red kidney beans
⅓ cup (75ml) Great Northern or lima beans, soaked
¼ cup (50ml) chick-peas, soaked
4 tablespoons olive oil
1 tablespoon vinegar
Salt and pepper
4 green onions, finely chopped
1 small green pepper, seeded and chopped
1 clove garlic, crushed
Small bunch of parsley, chopped

Cook the beans in unsalted water for 45 minutes or until tender. Drain well.

Toss the beans in the oil, vinegar, salt and pepper and add the remaining ingredients. Serve at room temperature or chilled.

Asinan   *Indonesia*

The salads of Indonesia, especially Java, are
particularly good. They make use of fruit, raw
and cooked vegetables, and slightly sweet and
sour dressings. If you can obtain udang kering
(dried shrimps) these can be added, fried and
crushed, with the dressing. The dried prawns
available at Chinese supermarkets can also be
used.

6 ounces (175g) bean sprouts (mung or soya)
1½ cups (375ml) very finely shredded cabbage
1 small cucumber, cut in julienne strips
1 tablespoon vinegar or lemon juice
1 teaspoon white or brown sugar
½ teaspoon salt
Pinch of chilli powder
2 tablespoons roasted peanuts, chopped

Toss the bean sprouts, cabbage and cucumber
with the lemon, sugar, salt and chilli powder.
Sprinkle with the peanuts and serve.

Aduki and Rice Salad (1)

A very fresh-looking salad which should be
served very cold.

¾ cup (150ml) long-grain rice
1 tablespoon oil
1 tablespoon lemon juice or vinegar
Salt and pepper
1 tablespoon chopped parsley
2 tablespoons chopped watercress
¼ cucumber, finely diced
1 small green pepper, seeded and finely diced
2 green onions, chopped, including the green
2 cups (500ml) sprouted aduki beans

Cook the rice in boiling salted water for 15 to 20
minutes until just soft. Drain well.

Stir the oil and lemon (or vinegar) into the rice
while it is still warm. Season and gently toss in
the other ingredients.

Aduki and Rice Salad (2)

Rice salads of all kinds seem to be popular, and make good party food. The addition of beans or lentils adds texture and colour as well as protein. Serve on its own before or after a light egg or fish dish, or to accompany cold meats.

½ cup (125ml) aduki beans, soaked
1 cup (250ml) long-grain rice
2 tablespoons oil
2 tablespoons lemon juice or wine vinegar
Salt and pepper
½ cup (125ml) frozen peas
½ cup (125ml) corn kernels
2 sticks celery, chopped
1 small onion, very finely chopped
½ small cucumber, diced
2 tablespoons raisins
2 tablespoons walnuts or pecans, chopped
Garnish:
Chopped parsley

Cover the aduki beans with water and cook for 30 to 45 minutes, taking care that they remain intact.

Boil the rice for 10 minutes in enough water to just cover it, cover with a clean teatowel and saucepan lid and leave on a low heat until the rice is fluffy and dry (5 to 10 minutes, depending on the rice).

Mix the oil and lemon or vinegar together, season highly and stir into the warmed rice.

Cook the frozen peas and add to the rice together with the corn kernels, the drained aduki beans, the chopped vegetables, raisins and nuts. Stir well and sprinkle with parsley. Serve chilled in a wide earthenware bowl or shallow terrine.

Damascus Bean Salad   *Middle East*

Ful medames with a spicy dressing. The tahina paste is widely available at delicatessens and some health food stores.

1 cup (250ml) fava beans (ful medames), soaked
3 tablespoons olive oil
2 tablespoons lemon juice
½ teaspoon turmeric
½ teaspoon cayenne pepper
6 tablespoons tahina paste
Salt
To serve:
2 hard-cooked eggs, quartered
Finely chopped parsley

Cook the ful medames in boiling, unsalted water for 1½ to 2 hours. Drain well.

Mix together the oil, lemon and spices, then beat in the tahina and add salt to taste. Stir carefully into the beans and allow to cool.

Arrange the hard-cooked eggs around the top of the salad and sprinkle with the parsley.

Ful Medames

Although this dish dates back to ancient Egypt, it is still popular enough to be considered as almost the national dish. Traditionally it is served with Hamine eggs – hard-cooked eggs coloured with onion skins – and the dressing is passed round separately. It can also be eaten warm.

1 cup (250ml) fava beans (ful medames), soaked
4 tablespoons olive oil
Juice of ½ lemon
2 cloves garlic, crushed

Cook the beans for 2 hours until very soft. Drain well and dress with the oil, lemon juice and garlic. Cool.

This can be served with sliced hard-cooked eggs and chopped parsley.

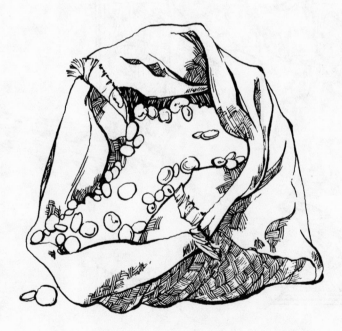

Eggs Condé   *France*

The eggs must be covered while baking or the yolks will set into a tough, leathery film.

1 cup (250ml) red kidney beans, soaked
1 onion, chopped
1 tablespoon butter
4 eggs
2 slices bacon
To serve:
Warm French bread

Cook the beans in just enough water to cover for 1 hour. Drain and reserve the liquid. Preheat the oven to 400°F (200°C).

Cook the onion in the butter until soft but not brown. Add the beans and 3 tablespoons of the cooking liquid. Stir well and cook for 5 minutes.

Purée to a smooth paste. Divide between 4 small ovenproof dishes and make a slight indentation in the top of each with a spoon. Break an egg into each hollow, cover with a lid or foil and bake for 9 to 12 minutes until the eggs are cooked but the yolk still soft.

Meanwhile, broil the bacon unitl very crisp and crumble over the cooked eggs. Serve hot with warm French bread.

Eggs Conti

A variation on the preceding recipe, using lentils instead of red beans.

1 cup (250ml) brown or green lentils
Salt
1 small onion, chopped
2 ounces (50g) ham, finely chopped
Salt and pepper
2 tablespoons butter
4 eggs

Cook the lentils in salted water with the onions for 40 minutes, then drain and reserve the liquid. Preheat the oven to 400° (200°C).

Purée the lentils, using a little cooking liquid to thin if necessary. Stir in the ham, salt, pepper and butter. Divide between 4 small ovenproof dishes, break an egg onto each, cover and cook for 8 to 12 minutes or until the eggs are lightly set.

Egg and Dal Curry

You can use a commercially prepared garam masala for this dish, but do not use curry powder.

Garam masala:
1 tablespoon ground coriander
2 teaspoons turmeric
1 teaspoon fenugreek
3 teaspoons ground cumin

2 tablespoons butter
1 onion, sliced
1 cup (250ml) masur dal or red lentils
4 hard-cooked eggs, quartered
To serve:
Rice or chappatis

Mix together the masala ingredients.

Melt the butter and fry the onion until golden brown. Stir in the masala and add the lentils. Lower the heat and stir the mixture for 5 minutes. Season with a pinch of salt and cover with water. Cook for 25 minutes until the lentils are tender and have absorbed the water.

Gently stir in the eggs. Warm through and serve.

Haricot Omelette   *Spain*

Romaine lettuce, shredded and tossed in a garlic-flavoured dressing and decorated with onion rings, anchovy fillets, a few olives and a chopped tomato, makes a good accompaniment.

1 cup (250ml) Great Northern beans, soaked
1 tablespoon olive oil
3 large or 4 medium eggs, beaten
Salt and pepper

Cook the beans in unsalted water for one hour. Drain.

Heat the oil in a large omelette pan and lightly stir in the beans. Add the beaten eggs, season and cook slowly for 5 to 8 minutes. Use a plate to turn the omelette and cook on the other side until lightly brown. Slide out onto a hot plate and serve with a salad.

Pasulj with Eggs

A delicious and nutritious lunch or supper dish.

1 cup (250ml) Great Northern beans, soaked
Salt and pepper
2 teaspoons olive oil
2 hard-cooked eggs, sliced
⅔ cup (150ml) sour cream
1 egg
½ cup (125ml) grated Gruyère, Emmenthal or Jarlsberg cheese

Cook the beans in unsalted water for 1 hour. Preheat the oven to 350°F (180°C). Drain the beans and sieve to a smooth purée.

Oil an ovenproof dish and spread in half of the purée. Arrange the hard-cooked eggs on top and spread with the remaining purée.

Beat the sour cream and egg together and pour over the bean purée. Sprinkle with the cheese. Bake for approximately 30 minutes and serve very hot.

A variation on the Scotch egg theme but far superior, using a meat and lentil coating for hard-cooked eggs, with a stuffing of prunes and walnuts. Best eaten cold, either whole or sliced, with a salad.

If you intend to serve these individually, you will need one egg per person, as below. For a less filling lunch, halve the recipe.

½ cup (125ml) channa dal or split yellow peas
1 pound (450g) lean lamb or beef, finely ground
2 tablespoons long-grain rice, cooked and cooled
½ teaspoon turmeric
½ teaspoon cinnamon
½ teaspoon grated nutmeg
½ teaspoon coriander
½ teaspoon salt
1 onion, finely chopped
1 teaspoon finely chopped parsley
4 prunes, soaked overnight in cold tea and chopped
¼ cup (50ml) walnuts or pecans, chopped
4 hard-cooked eggs
Oil
2 tablespoons tomato paste

Cook the channa dal or split peas for 45 minutes or until very soft. Drain and mash in a large bowl.

Beat the meat with a wooden spoon and add to the dal, beating well. Stir in the rice, spices, salt, onion and parsley. Mix well – the easiest way is to knead it with your hands – until it is a soft, workable paste.

Dust a pastry-board or large plate with flour. Divide the meat/dal paste into four and shape into flat squares. Mix the prunes and the nuts. Place one egg on each square, divide the prune and nut mixture between the four squares and arrange around the eggs. Shape the meat paste over each egg to enclose it and the stuffing, smoothing and pressing the meat to make a firm coating. Damp or lightly oiled hands stop this from being too messy a procedure.

Flour the rolls lightly and fry in a little hot oil, turning until they are evenly browned – cook gently to avoid breaking. Remove from pan.

Pour off most of the oil; stir the tomato paste into the water and pour into the pan. Bring to the boil, reduce and simmer very gently for 20 minutes, uncovered, allowing the sauce to thicken. Leave to cool, then serve with the eggs.

Another variation on the same theme, using a spicier coating and curried sauce. Bessan tastes very similar to channa dal, but its drier texture makes it better for coating.

½ pound (225g) lamb, minced (ground)
2 tablespoons minced onion
3 cloves garlic, crushed
1 teaspoon minced fresh ginger
½ teaspoon salt
½ teaspoon turmeric
¼ teaspoon cayenne pepper
2 tablespoons bessan (p.142), or 2 tablespoons channa dal, soaked overnight and crushed
1 egg, separated
½ teaspoon garam masala (p. 68)
4 hard-cooked eggs
Oil (for deep frying)
Sauce:
1 onion, chopped
2 cloves garlic, crushed
2 tablespoons ghee or oil
1 teaspoon fresh chopped ginger
½ teaspoon cumin
½ teaspoon turmeric
½ teaspoon ground coriander
1 teaspoon garam masala (p. 68)
2 tablespoons yoghurt
2 tomatoes, peeled and chopped
½ cup (125ml) water
To serve:
Cooked rice
1 tablespoon chopped parsley

Cover the meat with just enough water to moisten it, add the onion, garlic, ginger, salt, turmeric and cayenne and bring to the boil. Cover and simmer for 15 minutes.

Stir in the bessan or crushed dal. Cook, uncovered, until dry. Cool and transfer to a mixing bowl. Beat in the raw egg yolk and garam masala, beating to a smooth paste, or knead until smooth. Divide the paste into four. Using oiled or dampened hands, mould the paste around each egg, shaping carefully to give an even coating. Brush with a little egg white. Preheat the oven to 375°F (190°C).

To prepare the sauce: fry the onion and garlic in the ghee or oil for 3 minutes and stir in the ginger, spices, yoghurt, tomatoes and water. Cook gently, stirring, until the sauce thickens.

Heat the cooking oil to 350°F (180°C) and fry the koftas for ten minutes until golden brown on all sides. Turn carefully to avoid breaking. Arrange the drained koftas in a shallow fireproof dish, pour the sauce over and cover with foil. Cook for 20 minutes.

To serve, cut each kofta in half, arrange on the rice and pour the sauce around the eggs, but not on them. Sprinkle with parsley.

meat

Although no book on beans would be complete without this classic, I almost hesitate to give a recipe as the very name of the dish leads to dispute over ingredients and place(s) of origin. Cassoulet is basically a dish of beans cooked with a variety of meats which can include duck, pork, goose, lamb, sausages and mutton. In general, although individual family recipes do vary, the Cassoulet de Toulouse normally includes *confit d'oie* (preserved goose), pork, lamb and sausage. The Cassoulet de Castelnaudary, which claims to be original, uses only pork and sausages with the beans. It is not really a dish suitable for cooking in small quantities, or for small dainty appetites.

This recipe serves 8 to 10 people.

4 cups (1 litre) Great Northern beans, soaked
2 cups (500ml) chicken stock
¼ pound (100g) fresh pork rind, blanched and diced
1 Polish sausage, pricked
2 onions
Bouquet garni of 3 sprigs parsley, 2 bay leaves, 1 sprig thyme and 3 cloves garlic, tied together in muslin
½ pound (225g) salt pork in one piece
1 tablespoon oil (or rendered goose or pork fat)
1 pound (450g) lean pork, cubed
1 pound (450g) boned shoulder of lamb, cubed
2 tablespoons tomato paste
2 glasses dry white wine *or red?*
Salt and pepper
2 tablespoons chopped parsley
1 cup (250ml) dry white breadcrumbs

cloves pinch
carrots
barley ? added texture

Cook the beans in the stock for 30 minutes.

Meanwhile, simmer the pork rind in 1 cup (250ml) water for 20 minutes. Drain. Add the sausage, pork rind and one whole onion to the beans and cook for a further 30 minutes. Add the *bouquet garni* and the salt pork to the beans and cook very gently for another 40 minutes.

Meanwhile, heat the oil in a heavy pan and sear the pork cubes until brown all over. Remove with a slotted spoon and brown the cubed lamb in the same oil. Chop the remaining onion and add to the lamb. Cook gently until soft. Return the pork to the pan, stir in the tomato paste and wine and bring to the boil, stirring constantly. Cover and simmer gently for 1 hour, topping up the liquid with a little more stock or water if the mixture looks too dry. Remove the meat cubes.

Drain the beans, reserving the cooking liquid, and cut the salt pork and sausage into slices. Stir the beans into the meat juices, adding a little of the bean liquid and seasoning well. Simmer for 10 minutes, letting the beans absorb most of the liquid. Heat oven to 350°F (180°C).

Using a large cast-iron or enamelled casserole, layer the ingredients in the following way: a layer of beans, half the pork, lamb, rind, sausage and salt pork slices, then another layer of beans and the remaining meat, ending with a layer of salt pork and sliced sausage. Pour over the meat juices and add enough bean liquid to cover.

Mix the parsley and breadcrumbs together and sprinkle over the sausage, then top with a little rendered pork fat. Bake for up to 1½ hours. The crust which forms is usually broken up into the cassoulet after 30 minutes, allowing a second crust to form.

This is the national dish of Brazil and is really more of a party buffet based on a variety of smoked and fresh meats such as smoked tongue, dried beef and pickled pork, eggs with manioc meal sauce, rice, sliced oranges, shredded kale, hot chilli sauces and a dish of black beans in which the meats have been cooked. The meal takes its name from the Portuguese *feijão*, meaning bean.

I have assumed the proportions for 4 hungry adults, but by increasing the amount of rice and beans it could serve 8 to 10 for an unusual hot buffet meal. Although pig's ears, tails and feet are usually included, I have omitted them, with apologies to any expatriate Brazilians.

1 cup (250ml) black beans, soaked
6 ounces (175g) chuck beef in one piece
¼ pound (100g) salt pork in one piece
¼ pound (100g) slab bacon
½ pound (225g) pork spareribs
¼ pound (100g) chorizo or other smoked sausage
1 onion, chopped
1 clove garlic, chopped
1 tablespoon lard
1 7-ounce (198g) can tomatoes, drained
Salt and pepper
1 tablespoon Tabasco sauce (or less, according to taste)
2 oranges, peeled and sliced
Rice:
1 onion, sliced
1 tablespoon oil
1 cup (250ml) long-grain rice
1 tomato, peeled and chopped
Salt

Greens:
1 pound (450g) kale, shredded
1 tablespoon lard
Pepper sauce:
1 tablespoon Tabasco sauce
2 tablespoons lemon juice
1 clove garlic, crushed

Place the beans, beef, salt pork, bacon and spareribs in a heavy saucepan, cover with water and simmer for 1 hour. Add the chorizo, in one piece, and cook for another 30 minutes.

Meanwhile, prepare the rice: cook the onion in the oil until soft, add the rice and turn it in the oil for 2 minutes. Add the tomatoes and salt with enough water to cover. Cook gently, covered with a clean teatowel and the lid, for 15 minutes.

Remove the meats from the pan and drain the beans, reserving the cooking liquid. Cook the onion and garlic in the lard until soft, add the tomatoes, salt and pepper and Tabasco sauce. Add half the beans and mash them into the onion mixture. Add 3 tablespoons of the bean liquid and cook for 10 minutes until thick. Add to the remaining beans, together with the rest of the bean liquid, stirring and simmering until the mixture thickens. Keep warm.

Shred the kale and cook in 1 inch (2½cm) water for 3 minutes, boiling hard. Drain. Gently heat the lard and stir in the kale. Do not let it brown. Season and transfer to a heated dish.

Mix together the sauce ingredients.

Slice the meat and sausages and arrange on a plate. Put the beans, rice, kale, sauce and oranges in separate bowls and arrange on a table with the plate of meats.

This is really a dish for a large family occasion, needing large joints of beef and whole chickens. However, if cooked on a smaller scale the meats can be used in this dish and any left-overs eaten cold at other meals. The Spanish Cocido Madrileño is almost identical in every way, but usually includes smoked ham.

⅔ cup (150ml) chick-peas
1 pound (450g) beef rump roast or chuck beef (rolled and tied)
½ pound (225g) slab bacon or a ham hock
1 onion, thickly sliced
Salt and pepper
1 2-pound (900g) chicken, split in half
2 potatoes, peeled and quartered
2 carrots, peeled and sliced
2 turnips, peeled and quartered
½ cabbage, cut in chunks
¼ pound (100g) chorizo or any other smoked garlic sausage
½ cup (125ml) long-grain rice
Garnish:
Chopped parsley

Put the chick-peas, beef, bacon piece and onion in a large, heavy saucepan. Season, cover with cold water, bring to the boil and cook gently for 2 hours. Check that the meat is covered by water throughout the cooking. Add the chicken, potatoes, carrots and turnips and continue cooking for 20 minutes. Add the cabbage and the chorizo and check the seasoning.

Ladle about 1 cup (250ml) of the cooking liquid into a separate pan, bring to the boil and add the rice. Cook for 15 to 20 minutes until the rice is dry, and at the same time continue to cook the Cozida.

To serve, slice the beef, bacon and sausage and keep warm on a large heated dish. Bone the chicken, cut it into neat pieces and place on the serving dish. Strain the remaining ingredients and transfer the vegetables to the meat dish.

Serve the broth as a starter, followed by the dish of meats and vegetables accompanied by the rice in a separate dish, garnished with a little chopped parsley.

A very thick and filling stew.

1 tablespoon olive oil
1 large onion, chopped
1 red pepper, seeded and chopped
¾ pound (300g) chuck steak or stewing beef, cubed
2 teaspoons paprika
¼ teaspoon chilli powder
½ cup (125ml) Great Northern or butter beans, soaked
2 slices blanched bacon, diced
3 cups (750ml) beef stock
Salt and pepper
6 ounces (175g) chorizos, garlic or Polish sausage, sliced
¾ cup (175ml) corn kernels

Heat the oil in a heavy saucepan and cook the onion for 2 minutes, add the red pepper and continue cooking for another 3 minutes. Add the beef and brown all over.

Remove the pan from the heat and stir in the paprika and chilli. Add the beans and bacon and cover with the stock. Season with salt and pepper, bring to the boil and cover. Reduce the heat and simmer for 1½ hours.

Add the sausage and corn, and more liquid if the stew seems too dry. Simmer gently for 30 minutes and serve hot with fresh bread.

Bacon and Ham

Pasulj   *Bulgaria*

1 large onion, finely chopped
1 clove garlic, crushed
1 tablespoon oil or lard
2 teaspoons Hungarian paprika
1½ cups (375ml) Great Northern beans, soaked
¾ pound (350g) Canadian bacon or ham, cubed
¼ teaspoon cayenne pepper
1 cup (250ml) chicken or ham stock
Salt and pepper

Cook the onion and garlic in the oil or lard in a heavy saucepan, until soft and just beginning to colour.

Remove from the heat and stir in the paprika. Add the beans, bacon, cayenne and stock. Bring to the boil, cover, and simmer for 1½ to 2 hours. The liquid should be reduced to a thick sauce. Season to taste and serve hot.

Bacon, Beans and Noodles   *Yugoslavia*

An unsophisticated but very filling meal.

1 cup (250ml) Great Northern beans, soaked
½ pound (225g) slab bacon
1 cup (250ml) chicken stock
1 onion, chopped
2 cloves garlic, crushed
½ pound (225g) flat egg noodles
1 tablespoon chopped parsley
To serve:
Freshly grated Parmesan cheese

Cook the beans and the bacon in the stock for 1 hour. Remove the bacon and slice thinly.

Add the onion and garlic to the beans, bring to the boil and add the noodles. Cook briskly for 12 minutes then stir in the parsley, bacon slices, salt and pepper to taste.

Serve in deep bowls, with grated Parmesan and warm bread.

Beef

Chilli con Carne

Although most recipes for Chilli con Carne include red kidney beans, in its original Texan form it is essentially a fiery meat and chilli stew. Despite its Mexican name, it is definitely American and, with or without beans, is more a way of life than a mere dish. The following recipe (which does include beans) is hot, but the amount of chilli can be varied according to taste.

½ cup (125ml) red kidney or pinto beans, soaked
2 tablespoons oil
2 onions, chopped
1 clove garlic, crushed
1 pound (450g) chuck beef, coarsely ground or chopped
½ large red pepper (or 1 small one), seeded and finely chopped
½ teaspoon dried oregano
½ teaspoon ground cumin
1½ to 3 tablespoons chilli powder
4 tablespoons tomato paste
1 cup (250ml) beef stock
Salt and pepper
To serve:
Corn or flour tortillas

Cook the beans for 45 minutes, then drain.

Heat the oil in a heavy frying pan or saucepan and cook the onions and garlic gently until soft. Raise the heat and add the meat to the pan, stirring, until all traces of pink are gone. Stir in the red pepper, oregano, cumin and chilli. Stir in the tomato paste and stock. Season and bring to the boil. Cover and simmer for 1 hour 10 minutes. (This can be cooked in a covered casserole in the oven if more convenient).

Stir in the beans and cook for a further 20 minutes. Serve in deep bowls, with tortillas.

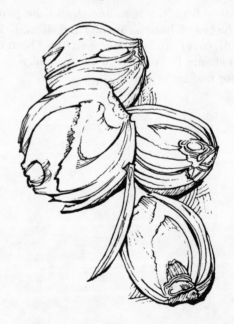

Beef and Haricot Casserole

1 cup (250ml) Great Northern or butter beans, soaked
1 pound (450g) chuck or braising beef, cubed
1 tablespoon flour
2 tablespoons oil or butter
2 onions, sliced
2 carrots, peeled and diced
1 bay leaf
1 8-ounce (225g) can tomatoes, drained and chopped
Salt and pepper
3 potatoes, peeled and thickly sliced
About 2 cups (500ml) beef stock

Cook the beans in unsalted water for 1 hour and drain. Coat the meat in flour and cook until brown in the oil or butter. Transfer to a casserole and add the beans.

Fry the onions and carrots in the oil for 2 to 3 minutes and add to the casserole. Add the bay leaf, tomatoes, salt and pepper and stir gently.

Lightly brown the potato slices in the oil (adding more oil if needed). Drain and arrange, overlapping, on the top of the casserole. Pour over enough stock to just cover. Cover with a lid and cook at 350°F (180°C) for 2 hours.

Pabellon Criollo   *Venezuela*

⅔ cup (150ml) black beans, soaked
2 cups (500ml) water
2 tablespoons oil
2 onions, chopped
1 clove garlic, crushed
Salt
1 small green pepper, seeded and chopped
4 tomatoes, peeled and chopped
½ teaspoon ground cumin
1 pound (450g) rump steak
¾ cup (150ml) long-grain rice

Cook the beans in the water for 1 hour. Heat ½ tablespoon oil and cook ⅓ of the onions, the garlic and a pinch of salt for 5 minutes over a very low heat. Add half the green pepper and cook for 3 minutes, stirring. Add 2 tablespoons of the bean liquid, stir well and pour the mixture into the beans. Cover and cook gently for 20 minutes.

Meanwhile, heat 1 tablespoon oil and cook ⅓ of the onions until soft but not brown. Add the tomatoes, cumin and a pinch of salt. Stir well, cover and cook for 20 minutes until thick.

Grill the steak until medium rare (or if using a less tender cut, sauté in a little oil). Cool and cut into thin strips. Add the beef to the tomatoes, stir well and cover to keep warm.

Heat the remaining oil in a saucepan and cook the remaining onion and green pepper. When soft but not coloured, add the rice, stirring well. Cover with water and a pinch of salt and bring to the boil. Reduce the heat and simmer for 20 minutes until the rice is dry and fluffy.

Heat through the meat and beans, separately. Arrange the rice, beans and meat decoratively on a large warm dish and serve.

Although in Orthodox Jewish homes this would be cooked overnight to avoid cooking on the Sabbath, I have given the cooking time for a normal meal. This is a delicious winter dish. The beans, prunes and barley should be soaked separately.

⅝ cup (150ml) Great Northern or butter beans, soaked
½ cup (125ml) pearl barley, soaked
¾ – 1 pound (350-450g) chuck, skirt or lean brisket in one piece
¾ cup (200ml) prunes, soaked overnight
4 potatoes, peeled
1 small onion, sliced
Salt and pepper
Dumpling:
½ cup (125ml) all-purpose flour
1 tablespoon minced onion
1 tablespoon chicken fat or lard
1 teaspoon chopped parsley
Salt and pepper
Approximately 4 tablespoons water

Mix together all the dumpling ingredients with enough water to make a soft dough.

Mix the beans and the barley together. In a large casserole or heavy saucepan arrange the following layers: barley/beans mixture, meat (in one piece), prunes, spoonfuls of dumpling mixture, and finally the potatoes and onion, seasoning each layer with a little salt and pepper. Add enough boiling water to just cover the potatoes. It is essential that the water is boiling, especially if you are going to cook it in a low oven.

Cover the pot with foil and a lid and either cook at 250°F (125°C) for 4 hours or bring to the boil and simmer on the top of the stove for 4 hours. Serve, slicing the meat, potatoes and dumpling into neat pieces.

If you can obtain plantains these can be added, peeled and sliced, with the potatoes.

2 slices blanched bacon, finely chopped
1 pound (450g) lean beef, cubed
1 large onion, chopped
½ cup (125ml) split yellow peas or pigeon peas
2 cups (500ml) beef stock
Salt and pepper
½ small fresh red chilli, or ½ teaspoon cayenne
  pepper
2 large potatoes, peeled and sliced
6 ounces (175g) sweet potatoes, peeled and sliced
½ cup (125ml) coconut milk (see note)
Dumplings
¼ cup (50ml) all-purpose flour
¼ cup (50ml) cornmeal or polenta
½ teaspoon baking powder
Pinch of salt
1 tablespoon butter
Approximately 1 tablespoon cold water

Cook the bacon slowly until the fat has run out. Remove the bacon and brown the beef in the bacon fat. Add the onion and cook, stirring, for 3 minutes. Add the split peas and stock and season well. Bring to the boil, cover and simmer for 1 hour.

Add the chilli or cayenne, potatoes, sweet potatoes and coconut milk and continue simmering for 10 minutes.

To prepare the dumplings, sift together the dry ingredients and rub in the butter until the mixture resembles fine breadcrumbs. Add the water, mixing it in quickly – you may need a little more water, but don't use more than a teaspoonful. Divide the mixture into 8 small dumplings and drop into the simmering stew.

Continue cooking, covered, for 10 to 15 minutes until the dumplings are cooked. Remove the chilli (if used) and serve hot in deep bowls.

Coconut Milk: Crack a fresh coconut and drain the liquid into a measuring cup. Dislodge the meat and cut into ½-inch (1cm) pieces. Add enough water to the coconut liquid to make ½ cup (125ml). Purée 1 cup (250ml) coconut chunks and the liquid in a blender. When smooth, place in a saucepan and heat very gently. Let steep for 30 minutes. Strain through 2 thicknesses of cheesecloth. Makes ½ cup (125ml).

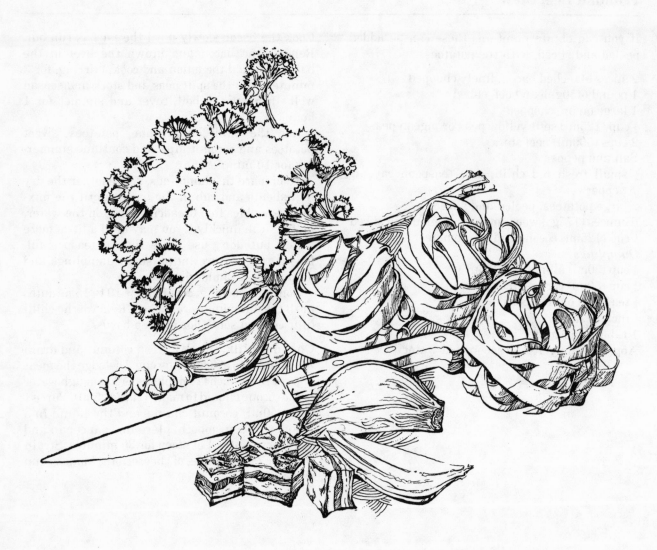

Beef and Vegetable Casserole *Greece*

Eggplant, green beans and red cabbage can be used instead of – or as well as – the vegetables suggested here.

1½ cups (375ml) Great Northern or fava beans, soaked
¾ pound (350g) stewing beef, cut into strips
1 onion, sliced
3 potatoes, peeled and sliced
2 tomatoes, peeled and chopped
1 carrot, scraped and sliced
2 zucchini, sliced
1½ cups (375ml) shredded cabbage
Salt and pepper
1 cup (250ml) beef stock
½ cup (125ml) dry white wine
To serve:
Crusty bread

Cook the beans until just tender in unsalted water, approximately 40 minutes, then drain.

Using a large earthenware or cast iron casserole, layer the meat, vegetables and beans, seasoning well and reserving a layer of potato for the top. Pour over the stock and wine. Cover and cook for 3 hours at 300°F (150°C), removing the lid for the last 30 minutes.

Dfina *Egypt*

A rich beef stew with a surprise – eggs in their shells. The particular character of this dish is obtained from the use of sorrel, for which there is no substitute.

Chick-peas are often used instead of haricot beans and traditionally the stew is enriched with a calf's foot. If you have an accommodating butcher this does improve the flavour, especially if chick-peas are used. Blanch the calf's foot and add with the meat.

1 pound (450g) stewing beef, cubed
1 cup (250ml) Great Northern beans, soaked
2 large onions, chopped
4 cloves garlic, crushed
4 eggs (in their shells)
1 teaspoon ground allspice
3 cups (750ml) water
1 pound (450g) sorrel
1 tablespoon olive oil
To serve:
Boiled rice

Put all the ingredients into a heavy saucepan or cast iron casserole, with the exception of the sorrel and oil. Bring to the boil, cover and simmer very gently for 4 hours, or bake in a very slow oven for the same length of time.

One hour before serving, blanch the sorrel, drain and chop. Cook gently in the oil for 3 minutes. Stir into the stew and continue the cooking, then serve piping hot – the eggs will be hard boiled but creamy.

Lamb

Gigot à la Bretonne *France*

Although this cannot be prepared with the pre-salé lamb as in its native Brittany, the method of cooking and use of haricot beans are the same, and make a good combination for a winter Sunday lunch. Shoulder of lamb can be roasted in the same way, in which case it will be Epaule à la Bretonne.

3 cloves garlic
A 1½ - 2 pound (675-900g) leg of lamb
1 cup (250ml) Great Northern beans, soaked
   overnight
2 small onions
1 carrot, peeled and quartered
1 bay leaf, 1 sprig thyme and 2 sprigs parsley, tied
   together
Pepper
Butter
1 tomato, peeled and chopped
Salt
To serve:
Brussels sprouts and chestnuts, braised in butter

Insert the cloves of garlic near the bone and roast the lamb at 400°F (200°C) for 15 minutes. Reduce the heat to 350°F (180°C) and continue cooking for 1 hour, basting occasionally.

Meanwhile, cook the beans in unsalted water with 1 of the onions, the carrot, herbs and pepper for 1 hour or until tender. Drain, reserving the liquid. Remove the herbs and vegetables.

Chop the remaining onion finely and cook in a little butter until it begins to colour. Stir in the tomato, beans and 2 tablespoons of the cooking liquid. Add salt to taste.

Remove the meat from the dish and pour off the fat, leaving any meat juices. Stir the beans into the juices and leave, covered, in a turned-off oven while you carve the meat. Serve the slices of meat with the bean mixture.

Absgooth (Abgooshth) *India*

This Moslem dish, although technically a soup, can be served in deep bowls as a stew.

1 pound (450g) lean lamb, cubed
A few lamb bones
1 cup (250ml) channa dal or split yellow peas, soaked
4 tomatoes, peeled and chopped
1 inch (2½cm) piece of cinnamon
1 teaspoon ground cardamom
4 whole cloves
1 large onion, chopped
1 eggplant, sliced and salted for 1 hour
2 tablespoons ghee or butter
Salt
Juice of 1 lime or lemon
To serve:
Warm bread

Simmer the lamb, bones, dal, tomatoes, spices and onion, with enough water to just cover, for 2 hours in a covered saucepan.

Wash and dry the eggplant and fry in the ghee or butter until deep brown. Drain.

Remove the bones from the stew and season with salt and the lime or lemon juice. Stir in the eggplant and serve with warm bread.

Khoresh Qormeh Sabzi *Iran*

2 tablespoons oil
2 onions, chopped
1 pound (450g) lean lamb, cubed
⅓ cup (75ml) red kidney beans, soaked
A few lamb bones
1 teaspoon ground fenugreek
1 teaspoon turmeric
½ inch (1¼cm) piece of cinnamon
2 tablespoons chopped parsley
1 teaspoon chopped mint
Salt and pepper
2 tablespoons lemon juice
To serve:
Rice

Heat 1 tablespoon oil in a heavy saucepan and cook 1 onion until golden. Add the lamb and stir until evenly browned. Cover with water, add the beans and bones and bring to the boil. Cover and simmer for 1 hour.

In a small frying pan heat the remaining oil and cook the second onion, the spices, parsley and mint for 3 minutes. Season with salt and pepper and transfer to the lamb and bean mixture. Continue cooking for 45 minutes, simmering gently and adding more water if the mixture becomes too dry.

Remove the bones, flavour with the lemon juice and serve with buttered rice.

This famous Parsee dish is a particular personal favourite – a rich stew of lamb, lentils, eggplant and spinach, so subtly spiced it makes the Anglo-Indian curry seem ostentatious. Breast of lamb is often used in India, but I prefer to use a leaner cut.

½ (125ml) tur dal
¼ cup (50ml) mung dal
¼ cup (50ml) masur dal
Masala:
1 tablespoon fresh ginger, peeled and chopped
1 clove garlic
4 whole cloves
6 peppercorns
1 inch (2½cm) piece of cinnamon
1 teaspoon whole black cumin
1 teaspoon coriander seeds
1 fresh red chilli, or ½ teaspoon cayenne pepper

2 onions, thinly sliced
4 tablespoons ghee or oil
1 large eggplant, cubed
¾ pound (350g) uncooked spinach, shredded
2 tomatoes, chopped
½ pound (225g) zucchini, cubed
½ to ¾ pound (225 to 350g) lean lamb, cubed
To serve:
Chopped parsley
Poppadums

Cook the dal in lightly salted water for 30 minutes.

Grind the masala ingredients to a thick paste in a blender or with pestle and mortar.

Meanwhile, cook the onions in 2 tablespoons ghee or oil until golden brown and soft. Remove half the onion slices for decoration and add the masala to the onions remaining in the pan. Cook, stirring over a low heat, for 3 minutes. Add the eggplant, spinach, tomato and zucchini, stir well, reduce the heat to its lowest point and cook, covered, for 20 minutes.

While the vegetables are cooking, lightly fry the lamb in the remaining ghee or oil then mix with the vegetables.

Drain the dal and mash with a little of the cooking liquid. Stir into the meat and vegetable mixture and continue cooking for 30 minutes on a very low heat until the meat is tender and the vegetables and dal have formed a thick spicy sauce.

Pour into deep bowls, sprinkle with the fried onions and parsley, and serve with poppadums.

Lamb and Chick-pea Pilav   *North Africa*

½ cup (125ml) chick-peas, soaked
3 tablespoons butter
1 onion, finely chopped
1 pound (450g) lamb, cubed
1 teaspoon ground cumin
1½ cups (375ml) long-grain rice
Salt and pepper

Cook the chick-peas for 1 hour. Drain and reserve the liquid.

Heat the butter in a heavy saucepan and fry the onion until golden. Raise the heat and add the lamb, stirring until it is well browned on all sides. Add the cumin and the chick-peas; stir for 1 minute. Push these cooked ingredients to the side of the pan and add the rice, stirring until the grains become translucent.

Pour over enough chick-pea liquid to barely cover the rice. Stir well, season and simmer gently for 20 minutes, until the rice is dry and has absorbed the liquid. Serve very hot.

Dizi   *Iran*

½ cup (125ml) chick-peas, soaked
½ cup (125ml) fava beans (ful medames), soaked
1 pound (450g) lean lamb, cubed
A few lamb bones, if available
2 cups (500ml) lamb or chicken stock, or water
Salt and pepper
2 tomatoes, peeled and chopped
1 large onion, sliced
2 tablespoons lemon juice
Pinch of turmeric

Place the chick-peas and fava beans in a heavy saucepan with the meat, bones, and liquid. Season with salt and pepper, bring to the boil, cover and simmer for 45 minutes to 1 hour.

Add the remaining ingredients and cook gently for a further 40 minutes, making sure the stew does not get too dry.

Strain the liquid and keep warm. Discard the bones. Mash the chick-peas and beans coarsely.

Serve the liquid as a soup, followed by the meat and crushed beans. Serve with rice or bread if hunger necessitates.

Lamb and Haricot Paprika   *North Africa*

1 tablespoon oil
1 onion, sliced
1 red pepper, peeled, seeded and sliced
1 pound (450g) lamb or mutton, diced
2 to 3 teaspoons paprika
1 tablespoon flour
Salt and pepper
½ cup (125ml) Great Northern beans, soaked
2 tomatoes, peeled and chopped
3 potatoes, peeled and thickly sliced

Heat the oil and fry the onion, pepper and meat for 5 minutes. Transfer to a casserole with a slotted spoon.

Remove the frying pan from the heat, stir in the paprika and flour, return to the heat and stir in 1½ cups (375ml) water. Season well and add the beans and tomatoes. Bring to the boil and pour into the casserole.

Arrange the potato slices on top, cover with a lid or foil and bake at 350°F (180°C) for 2 hours.

Lamb and Lentil Stew   *Sweden*

As with most lamb stews or casseroles, this dish is best cooked ahead of time and allowed to cool, so that the fat can be removed.

1 tablespoon butter or oil
1 onion, sliced
1 pound (450g) lean lamb, cubed
2 cups (500ml) beef stock
Salt and pepper
1 bay leaf
1 sprig marjoram
1 cup (250ml) red or brown lentils, soaked for 3
  hours
Garnish:
Chopped parsley

Heat the oil in a saucepan and cook the onion gently until soft but not coloured. Remove and set aside.

Raise the heat and brown the lamb well, stirring to avoid sticking. Return the onions, pour over enough stock to cover, season and simmer for 30 minutes.

Add the herbs and lentils and continue simmering for another 45 minutes.

Allow to cool and remove the fat from the top. Bring slowly to the boil and remove the herbs. Sprinkle with chopped parsley and serve hot.

Most khichiris are spiced dal and rice stews, although they sometimes contain meat (as in this case) or, in their Anglo-Indian form – when they are known as kedgeree – fish and eggs.

2 tablespoons ghee or butter
1 onion, sliced
1 clove garlic, crushed
1 inch (2½cm) fresh ginger, peeled and grated
½ teaspoon ground cloves
½ teaspoon cinnamon
1 teaspoon ground cardamom
1 teaspoon turmeric
1 teaspoon ground coriander
1 dry chilli, or ¼ to ½ teaspoon powdered chillis
¾ cup (175ml) long-grain rice, preferably basmati
¾ cup (175ml) masur dal
¾ pound (350g) lean lamb or mutton, cubed
1 teaspoon salt
2 tablespoons butter
Raita:
1 onion, sliced
½ cup (125ml) plain yoghurt
Lemon juice to taste

Melt half the ghee and fry the onion until brown. Remove and set aside.

Add the garlic, ginger and spices to the pan and cook, stirring, for 2 minutes. Add the rice and dal, stirring well to coat with the spices.

In a separate pan, fry the meat in the remaining ghee until well browned. Add to the rice and lentils, together with the salt, butter and enough water to cover. Bring to the boil, cover and simmer for 1½ to 2 hours or until the meat is tender. Add more water if the mixture seems too dry.

Mix together the raita ingredients and serve with the khichiri.

Shami Kebab

Spiced meat and channa dal croquettes. Grind the meat twice using the finest blade of a meat grinder. The meat must be a paste otherwise the kebabs may break as they cook. Some recipes suggest boiling the dal and meat together, but this results in a poorer-textured kebab.

1 onion, minced
1 clove garlic, crushed
Ghee or butter
1 pound (450g) lamb, very finely ground and
    pounded to a paste
4 tablespoons water
1 tablespoon crushed fresh ginger
½ tablespoon ground cumin
½ teaspoon ground cardamom
½ tablespoon ground coriander
½ teaspoon salt
½ teaspoon pepper
¼ cup (50ml) channa dal, lightly roasted in a dry
    pan and ground or finely crushed
1 egg white, lightly beaten
Oil
To serve:
Indian bread (nan or roti)
Chutneys
Salads

Cook the onion and garlic for 2 minutes in the ghee or butter. Add the meat, stir and cook, covered, for 5 minutes, shaking the pan occasionally. Add the water to the meat and cook, stirring, until the water has been absorbed.

Transfer the meat to a bowl and stir in the ginger, spices, salt, pepper, channa dal and the egg white. Knead with your fingers (or beat with a wooden spoon) until the mixture is very soft and evenly mixed. Form into flat round cakes about 2 inches (5cm) in diameter.

Heat the oil in a large frying pan and fry the cakes for 5 to 10 minutes on both sides until golden.

A delicious spiced casserole of lamb and a variety of beans. Although breast of lamb is traditionally used I prefer a leaner cut. This is a simplified version of the original recipe, which suggests crushing or puréeing the beans before returning the meat.

1 small onion, minced
2 tablespoons butter
1 pound (450g) lean lamb, cubed
½ cup (125ml) red kidney beans, soaked
¼ cup (50ml) Great Northern beans, soaked
½ cup (125ml) black eyed peas
¼ cup (50ml) split peas or green lentils
2 tomatoes, peeled and chopped
1 teaspoon turmeric
½ teaspoon mace
½ teaspoon pepper
½ teaspoon paprika
½ teaspoon ground cardamom
Pinch of salt
1 bay leaf
1 tablespoon lemon juice
To serve:
Chappatis or poppadums

Cook the onion in half the butter until soft, then add the meat, beans and tomatoes and enough water to just cover. Cook gently for 20 minutes. Remove the meat with a perforated spoon and dry on paper towels, leaving the beans to continue cooking.

Cook the meat in the remaining butter until browned, stirring in the spices. Cook for 1 minute then add the green pepper and bay leaf. Stir this mixture into the beans. Continue cooking for 30 minutes or until the meat is tender.

Sprinkle with lemon juice and serve with chappatis or poppadums.

Pork

Beans and Sparerib Stew   *Yugoslavia*

1½ cups (375ml) Great Northern beans, soaked
½ pound (225g) meaty pork spareribs, chopped into
  large pieces
2 onions, sliced
2 tablespoons butter
2 cloves garlic, crushed
Salt and pepper
¼ teaspoon chilli powder or cayenne pepper
4 tomatoes, peeled and chopped
1 teaspoon Hungarian paprika
To serve:
Crusty bread

Put the beans and pork in a heavy saucepan and
just cover with water. Simmer, partly covered, for
1 to 1½ hours.

Cook the onion in half the butter until soft but
not coloured, then add the garlic, salt, pepper and
chilli. Stir well and add the tomatoes. Cook gently
for 5 minutes then add to the beans, stirring to
disperse the onions and tomatoes.

Melt the remaining butter and mix with the
paprika. Add 1 tablespoon of the bean cooking
liquid, stir well and add to the stew. Serve in deep
bowls with fresh crusty bread and follow with a
simple salad.

Bulgarian Bean Stew   *Bulgaria*

This can be eaten on its own or with noodles.

1½ cups (375ml) Great Northern beans, soaked
1 tablespoon lard or oil
½ to ¾ pound (225 to 350g) lean pork, cubed
2 onions, sliced
½ teaspoon chilli powder
1 cup (250ml) chicken stock
2 tomatoes, peeled and chopped
2 tablespoons tomato paste
Salt and pepper
Garnish:
3 sprigs mint, finely chopped

Cook the beans in unsalted water for 1 hour.

Meanwhile, heat the lard and brown the pork
cubes on all sides. Add the onion and chilli powder
and cook, stirring, for 5 minutes. Add the stock and
bring to the boil, simmering for 40 minutes.

Drain the beans and add to the meat, together
with a little of the bean cooking liquid if the
mixture seems too dry. Add the tomatoes, tomato
paste, salt and pepper. Cook for 15 minutes and
serve, sprinkled with the chopped mint.

Poultry and Rabbit

Chicken, Lentil and Eggplant Khoresh

You can use zucchini in place of the eggplant.

2 tablespoons olive oil
1 onion, chopped
1 small roasting chicken, cut up, or 4 chicken
  quarters
¼ cup (50ml) brown lentils, soaked
Salt and pepper
½ teaspoon turmeric
½ teaspoon ground cinnamon
2 tablespoons tomato paste
Juice of 1 lemon
1 eggplant, sliced
2 cloves garlic, crushed
To serve:
Steamed rice

Heat 1 tablespoon olive oil and cook the onion gently for 10 minutes.

Skin the chicken and add to the pan, browning on all sides. Add the lentils to the chicken, stirring to coat them with oil. Add enough cold water to barely cover the meat and season with the salt, pepper and spices. Simmer gently for 1½ hours.

Remove the chicken, cut the meat from the bones and return it to the pan. Stir in the tomato paste and lemon juice.

Fry the eggplant slices in the remaining oil in a separate pan until golden brown. Add the garlic and stir until it begins to change colour, then stir the eggplant and garlic into the chicken mixture. Cook for a further 20 minutes and serve hot with steamed rice.

6 tablespoons butter
1 small chicken, cut up, or 4 chicken quarters
1 onion, sliced
½ teaspoon paprika
1 teaspoon ground cumin
Salt
½ teaspoon saffron, steeped in 1 tablespoon hot
  water
2 cups (500ml) chicken stock
½ cup (125ml) chick-peas, soaked
1 cup (250ml) long-grain rice
Garnish:
2 tablespoons chopped parsley
Juice of 1 lemon

Heat 4 tablespoons of the butter in a heavy saucepan and sauté the chicken pieces until they begin to brown. Add the onion, paprika, cumin and salt and continue cooking gently for 5 minutes. Add the saffron and its soaking water (you may prefer to discard the saffron stamens themselves, I leave them in), chicken stock and chick-peas. Cover and simmer for 1½ to 2 hours, stirring occasionally.

Cook the rice for 15 to 20 minutes, adding the remaining butter to the water. When dry and fluffy arrange the rice on a serving dish. Place the chicken on top and spoon over the sauce. Sprinkle with parsley and plenty of lemon juice.

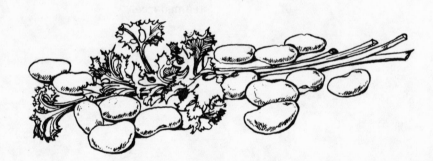

¾ cup (175ml) chick-peas, soaked
2 pounds (900g) rabbit, cut into 8 pieces
1½ tablespoons salted flour
2 tablespoons olive oil
1 onion, finely chopped
1 clove garlic, crushed
1 bay leaf
Salt and pepper
¼ pound (100g) chorizo or any smoked garlic
    sausage, sliced
1 cup (250ml) long-grain rice
3 eggs, beaten

Cook the chick-peas for 1 hour.

Lightly coat the rabbit pieces with the flour. Heat the oil and brown the rabbit on all sides. Remove from the pan and add the onion and garlic. Cook gently until soft but not brown. Add the meat, onion and garlic to the chick-peas.

Ladle 2 to 3 tablespoons of the chick-pea liquid into the pan used to fry the meat and bring to the boil, scraping any browned bits into the liquid. Pour back into the chick-peas. Add the bay leaf and a little salt and pepper.

If using chorizo, prick with a knife and boil for 5 minutes in a separate pan. Slice and add to the stew. (If other garlic sausage is used, don't add it until 5 minutes after the rice is added.) Continue cooking for 40 minutes.

Add the rice and cook briskly for 15 to 20 minutes until the rice is cooked and has absorbed most of the cooking liquid.

Heat the oven to 375°F (190°C). Pour the entire contents of the saucepan into an ovenproof serving dish.

Pour the eggs over the top and bake for 10 to 12 minutes or until the eggs are cooked and beginning to brown. Serve hot.

Spiced Red Beans   *Portugal*

Good either on its own or as a side dish with roast pork or grilled pork chops.

1 cup (250ml) red kidney beans, soaked
1 ham bone
Salt
¼ pound (100g) bacon, diced
1 onion, chopped
3 cloves garlic, crushed
1 small green pepper, seeded and diced
½ pound (225g) tomatoes, peeled, seeded and
   chopped
1 tablespoon tomato paste
Pepper
¼ teaspoon ground cinnamon
1 teaspoon ground cumin
¼ teaspoon chilli powder

Cook the beans with the ham bone for 1 hour in lightly salted water.

Meanwhile, lightly fry the bacon until all the fat has run out. Add the onion, garlic and pepper, stir and cook gently for 10 minutes. Add the tomatoes, tomato paste, salt, pepper and spices.

Drain the beans, reserve the liquid and discard the ham bone. Stir the beans into the cooked vegetables, pour over enough cooking liquid to just cover and simmer for 45 minutes. Serve hot.

Tavcé   *Bulgaria*

1 cup (250ml) red kidney beans, soaked
2 tablespoons oil
2 large onions, sliced
1½ tablespoons paprika
Salt and pepper
1 bay leaf

Cook the beans for 20 minutes. Drain and reserve the liquid.

Heat 1 tablespoon of oil and cook the onion gently until brown. Remove from the heat and stir in the paprika.

Use the remaining oil to grease a casserole. Heat the oven to 300°F (150°C). Put half the beans in a layer in the bottom of the casserole, season, cover with a layer of onions and top with the remaining beans. Season and bury the bay leaf in the centre.

Pour over enough of the bean liquid to barely cover the top layer. Cover with a lid or foil and bake for 1 hour, removing the lid for the last 10 minutes. Serve hot.

Red Beans in Wine

A baked bean dish with a difference. Any full red wine can be used – I can recommend the use of home-made elderberry.

1½ cups (375ml) red kidney beans, soaked
1 tablespoon olive oil
1 onion, chopped
2 cloves garlic, chopped
1 small red pepper, seeded and sliced
1 cup (250ml) red wine
1 tablespoon tomato paste
Salt and pepper
Butter

Cook the beans for 20 to 30 minutes in unsalted water until just tender. Drain. Pre-heat the oven to 350°F (180°C).

Heat the oil in a saucepan and cook the onion, garlic and pepper until soft. Add the beans, wine, tomato paste and bay leaf, and season well.

Transfer to an ovenproof dish, cover and bake for 45 minutes. Stir in a little butter and serve hot.

Frijoles Refritos   *Mexico*

Re-fried beans is the literal translation, a method of cooking which developed after the introduction of the pig to South America by the Conquistadors. In Mexico this dish would include a very hot tiny chilli known as tepin or pequin, but dried red chilli, chilli powder or cayenne pepper can be substituted.

1 cup (250ml) red kidney or pinto beans, soaked
½–1 teaspoon dried, crumbled chilli, or chilli powder to taste
2 onions, chopped
1 clove garlic, crushed
Pepper
6 tablespoons lard or oil
Salt
2 tomatoes, peeled, seeded and chopped
To serve:
Tortillas or warm bread

Put the beans, chilli, 1 onion, garlic and pepper into a saucepan. Cover with water, bring to the boil and simmer for 1½ hours.

Uncover the beans and boil rapidly until they are practically dry. Stir in 2 tablespoons lard and season with salt.

Heat the remaining lard or oil and fry the remaining onion, stirring until soft and lightly browned. Add the tomatoes and cook gently for 5 minutes.

Remove the onion and tomato from the pan and add the beans, crushing them coarsely with a fork or wooden spoon. Stir over a moderate heat for 2 minutes. Return the tomato mixture to the pan, reduce the heat and cook, stirring, for a further 5 minutes. Serve on its own with tortillas or bread.

Spiced Red Beans   *India*

Serve this as the centre of a vegetarian meal, with rice or chappatis, a selection of chutneys, raitas (yoghurt-based side dishes), and a cooked vegetable dish or curry.

1½ cups (375ml) red kidney beans, soaked
2 tablespoons oil
1 small onion, chopped
2 cloves garlic, chopped
1 inch (2½cm) fresh ginger, peeled, chopped and crushed
1½ teaspoons turmeric
Pinch of cayenne pepper
½ tablespoon ground cumin
1 tablespoon ground coriander
1 teaspoon garam masala (p. 68)
2 tomatoes, chopped
1 teaspoon salt
To serve:
1 tablespoon melted ghee or butter

Cook the beans in unsalted water for 45 to 60 minutes, or until just tender. Drain, reserving the liquid.

Heat the oil in a saucepan and cook the onion, garlic and ginger for about 8 minutes until soft and just beginning to brown. Stir in the spices and fry, stirring, for a few seconds. Add a little of the bean liquid and cook quickly until dry. Add the tomatoes, beans, remaining liquid and the salt. Bring to the boil, cover and simmer for 15 minutes.

Serve in a large heated bowl with the melted ghee or butter poured over.

Spiced Lobia Dal   *India*

A very good way of serving black eyed peas and one which can be used for soya beans.

1 cup (250ml) black eyed peas, soaked
1 dried red chilli, or ¼-½ teaspoon chilli powder or cayenne pepper
2 tablespoons ground coriander
½ cup (125ml) plain yoghurt
2 cloves garlic, crushed
1 tablespoon lemon juice
Salt and pepper
1½ tablespoons garam masala (p. 68)
Garnish:
1 teaspoon chopped mint

Cook the peas for 2 hours in unsalted water with the chilli and coriander. Drain and remove chilli. Mix together the yoghurt, garlic, lemon juice and salt and pepper to taste. Stir in the garam masala and add the yoghurt mixture to the hot peas. Serve sprinkled with mint.

Uppama   *India*

An unusual dish where lentils and fresh vegetables are cooked in a thickened spicy sauce. The chilli can be added as 2 or 3 large pieces to be removed later, if their inclusion makes the sauce too hot for your taste.

2 tablespoons oil
1 teaspoon whole black mustard seeds
1 small onion, chopped
½ cup (125ml) urd dal
1 small fresh green chilli, seeded and finely chopped, or ¼ to ½ teaspoon powdered chillis
⅓ cup (75ml) potato flour (potato starch) or cornstarch
2 cups (500ml) water
3 tomatoes, chopped
3 green onions, chopped, including the tops
1 carrot, peeled and finely diced
1½ teaspoons salt
Juice of ¼ lemon
Garnish:
Chopped parsley

Heat the oil in a heavy saucepan and add the mustard seeds (which will pop and splutter), the onion and urd dal and cook for 5 or 6 minutes. When lightly golden, stir in the chilli and cook for 1 minute.

Add the potato flour in a slow trickle, stirring continuously until the flour begins to colour. Gradually stir in the water and bring to the boil, still stirring. Add the vegetables and salt, cover and cook gently for 12 to 15 minutes.

Stir in the lemon juice and serve sprinkled with parsley.

Lentil and Potato Pie

1½ cups (375ml) lentils, soaked for 3 hours
Salt
4 large potatoes
Vegetable or meat stock
2 onions, sliced
2 tablespoons butter or oil
2 tablespoons flour
Pepper
1 teaspoon dried thyme
1 tablespoon chopped parsley
1 teaspoon chopped fresh basil, or ½ teaspoon dried
½ teaspoon Kitchen Bouquet

Cook the lentils in lightly salted water for 40 minutes or until very soft. Boil the potatoes in their skins and mash them when cooked.

Drain the lentils and reserve the liquid, making it up to 2 cups (500ml) with the stock.

Cook the onions in 1½ tablespoons butter or oil until golden. Stir in the flour, then gradually add the reserved liquid, cooking gently until a medium-thick sauce results – you may not need all the liquid for this. Season, add the herbs, Kitchen Bouquet and lentils. Cook gently for 10 minutes until thick.

Transfer to an ovenproof dish. Cover with the mashed potato, dot with the remaining butter and bake at 375°F (190°C) for 1 hour. Serve with a colourful fresh vegetable.

Madras Spiced Lentils   *India*

1 cup (250ml) masur dal or red lentils
1 teaspoon chilli powder
2 cloves garlic, crushed
1 onion, very finely chopped
1 inch (2½cm) fresh ginger, peeled and chopped
1 teaspoon salt
2 tablespoons ghee or melted butter
2 onions, sliced

Dry roast the dal in a heavy saucepan over a high heat for 1 minute, shaking the pan to avoid burning.

Lower the heat and add the chilli powder, garlic, chopped onion, ginger and salt. Cover with water, bring to the boil and simmer gently for 40 minutes.

Heat the ghee and fry the sliced onions until deep brown. Drain the dal, mash or purée them and serve in a heated bowl, sprinkled with the fried onions.

Buttered Lentils   *Britain*

A delicious and simple way to prepare brown lentils, good on its own or with meat dishes. The quantities given are for four to six large helpings.

3 tablespoons unsalted (sweet) butter
1 onion, chopped
1 clove garlic, chopped
1¼ cups (300ml) brown lentils, soaked for 30 minutes
Salt and pepper
Lemon juice

Heat 1 tablespoon of the butter in a saucepan and gently fry the onion and garlic until soft and beginning to colour.

Add the lentils, cover with cold water, season and bring to the boil. Partly cover the pan and simmer for 45 minutes to 1 hour, or until the lentils are soft and the liquid almost absorbed.

Stir in the remaining butter. Check the seasoning and add lemon juice to taste.

Lentils with Anchovy Butter   *France*

1 cup (250ml) brown or green lentils, soaked 1 hour
Salt
4 anchovy fillets, drained and chopped
4 tablespoons unsalted (sweet) butter, at room
    temperature
1½ tablespoons oil
1 clove garlic, crushed
Pepper
Garnish:
Chopped parsley

Cook the lentils in lightly salted water for 30 to
40 minutes, until tender but still whole.

Pound together the anchovies and the butter
until smooth.

Heat the oil and cook the garlic until just
beginning to colour. Stir into the lentils, season
with pepper and transfer to a warm dish.

Put the anchovy butter on top and stir gently as
it melts. Sprinkle with parsley and serve hot.

Musakka'a   *North Africa*

This Arab dish, not to be confused with Mous-
saka, is a delicious and economical main meal. It
is traditionally served at room temperature, but
is also very good hot, or cold as a salad.

1½ cups (375ml) chick-peas, soaked
1 large or 2 small eggplants, cubed
3 tablespoons olive oil
2 large mild onions, thickly sliced
Salt and pepper
1 pound (450g) tomatoes, peeled and chopped
To serve:
Pitta bread

Cook the chick-peas in unsalted water for 1 hour,
then drain. Meanwhile, heat the oven to 400°F
(200°C).

In a frying pan heat the oil and cook the eggplant
cubes until lightly browned, then transfer to an
ovenproof serving dish or casserole.

Cook the onions in the oil left in the pan for 10
minutes or until soft and golden. Cover the
eggplant with the onions, season well and pour
over a little oil. Cover with the chick-peas and cover
them, in turn, with the tomatoes. Season and pour
in enough boiling water to barely cover. Bake for
45 minutes, uncovered, checking occasionally that
it has not dried up.

Spiced Urd Dal   *India*

A simple way of preparing urd dal, in which they remain whole.

1 cup (250ml) urd dal
3 cups (750ml) water
Salt
1 teaspoon turmeric
½ teaspoon cayenne pepper
2 tablespoons oil
1 small onion, chopped
1 inch (2½cm) fresh ginger, peeled, chopped and crushed
1½ teaspoons ground coriander
1 teaspoon ground cumin
Garnish:
½ teaspoon garam masala (p. 68)
1 tablespoon chopped parsley
2 tablespoons melted ghee or butter

Cook the dal in the water with the salt, turmeric and cayenne over a gentle heat for 40 to 45 minutes or until the dal is soft but not breaking apart.

In a small frying pan heat the oil and cook the onion and ginger for 5 minutes. Add the coriander and cumin and continue cooking, stirring continuously, for 6 or 7 minutes. Stir into the dal and heat through for a few minutes.

Pour into a bowl and sprinkle with the garam masala, chopped parsley and melted ghee or butter.

Curried Urd Dal   *India*

1 cup (250ml) urd dal, soaked
2 tablespoons butter
1 level teaspoon salt
½ teaspoon turmeric
1 bay leaf
1 inch (2½cm) fresh ginger, peeled and thinly sliced
¼ teaspoon cayenne pepper
2 tablespoons parsley, finely chopped
4 cardamom pods
2 cloves garlic
1 teaspoon cumin seeds

Cook the dal in unsalted water for 40 minutes until they are tender and most of the water has been absorbed.

Remove from the heat and stir in the butter, salt, turmeric, bay leaf, ginger, cayenne and parsley. Crush 2 of the cardamom pods and add, then either simmer for 1 hour on the lowest possible heat or cook, covered, in a very slow oven for 1 hour. Check that the dal does not boil dry.

Just before serving, remove the skins from the 2 remaining cardamom pods and crush with the cumin and garlic. Stir into the dal. Leave, covered, off the heat for 5 minutes, then serve.

Tur Dal Masallah   _India_

1 cup (250ml) tur dal (channa dal) _split pea yellow_
1 teaspoon salt
1 teaspoon turmeric
1 inch (2½cm) fresh ginger, peeled and chopped
2 onions, sliced
2 zucchini, diced
½ inch (1¼cm) piece of cinnamon
4 cardamom pods, skinned
½ teaspoon pepper
½ teaspoon powdered chillis or cayenne pepper
2 tablespoon chopped parsley
2 cloves garlic, crushed
1 tablespoon ghee or butter
To serve:
Boiled rice
Tomato and onion salad

Put all the ingredients except the garlic and butter into a saucepan, cover with water and bring to the boil. Cover and simmer for 45 minutes until the dal and the vegetables are tender. Drain.

Gently cook the garlic in the ghee or butter. Add to the lentils and serve hot with rice and a tomato and onion salad.

Buttered Urd Dal   _India_

Buttered lentils of this type are sometimes left cooking overnight or for at least 6 to 7 hours to produce a very smooth buttery purée. The amount of butter may seem rather extravagant, but it does give the dish its characteristic flavour and texture.

1 cup (250 ml) urd dal, soaked for 3 hours
½ cup (125ml) butter
Seeds from 3 cardamom pods
Salt and pepper
2 tablespoons plain yoghurt
To serve:
Pitta bread, chutneys and yoghurt

Place the urd dal in a heavy pan with the butter, cardamom, salt and pepper. Barely cover with water, bring to the boil, cover and reduce heat. Simmer for 2 to 3 hours, adding a little more water if it becomes too dry. Alternatively, cook in a covered casserole in the oven at 300°F (150°C) for 3 hours.

Stir in the yoghurt, check the seasoning and serve.

Spiced White Urd Dal   *India*

1 cup (250ml) white urd dal, soaked overnight
4 tablespoons unsalted (sweet) butter
Seeds from 3 cardamom pods
½ teaspoon salt
Pepper
To serve:
1 teaspoon chopped parsley
1 teaspoon chopped chives

Put the dal, butter, cardamom, salt and pepper into a saucepan. Barely cover with water. Bring to the boil and cover, reduce the heat to very low and cook gently for 1½ hours, when the dal should be soft but still whole.
　Serve hot, sprinkled with the chopped herbs.

Bengali Masur Dal
India

Serve with a curried vegetable or meat dish.

1 cup (250ml) masur dal, or ordinary red lentils, washed and well dried
1 small onion, peeled and grated
½ inch (1¼cm) fresh ginger, peeled and grated
¼ teaspoon turmeric
½ teaspoon chilli powder
1 clove garlic, crushed
1 teaspoon salt
2 tablespoons tamarind water (or lemon juice)
3 tablespoons ghee, melted butter or oil
2 large onions, sliced

Put the dried dal into a heavy pan and dry roast over a moderate heat for ½ minute. Add the grated onion, ginger, spices, garlic salt, tamarind water and enough cold water to cover. Bring to the boil, cover and simmer for 1 to 1½ hours, the longer and slower the better.
　Heat the ghee and fry the sliced onions until deep brown. Lightly crush the lentils but do not purée, add the fried onions, stir well and cover. Remove from the heat and leave for 5 minutes before serving.

Spiced Mung Dal   *India*

In this recipe the mung beans are left whole and served as part of a vegetarian meal.

1 cup (250ml) mung dal (green or brown), soaked
    for 30 minutes
Salt
Pinch of cayenne pepper
1 teaspoon turmeric
2 tablespoons ghee, butter or oil
1 onion, very thinly sliced
1 clove garlic, crushed
1 teaspoon crushed cumin seeds
Garnish:
Chopped parsley

Put the dal in a saucepan and cover with water. Add the salt, cayenne and turmeric and bring to the boil, then simmer for 30 minutes until the dal is soft and the liquid greatly reduced.

Heat the ghee, butter or oil and fry the onion, garlic and cumin until the onion is soft and golden. Stir into the cooked dal and serve sprinkled with chopped parsley.

Curried Dal   *Ceylon*

1 cup (250ml) masur dal (red) lentils
1 onion, chopped
2 cloves garlic, crushed
½ teaspoon turmeric
1½ cups (375ml) water
1 cup (250ml)   coconut milk (p. 81)
½ teaspoon pepper
1 teaspoon ground cumin
¼ to ½ teaspoon chilli powder
Salt

Cook the dal with the onion, garlic and turmeric in the water for 40 minutes.

Add the coconut milk, pepper, cumin, chilli powder and a little salt. Bring almost to the boil and serve with chappatis and chutneys.

There are many versions of this dish, but all are basically a hot and spicy mixture of soft lentils and crisply cooked vegetables. Other dals and vegetables can be used. Serve as a main course.

1 cup (250ml) tur dal
3 tablespoons oil
Salt
1 tablespoon tamarind infused in ½ cup (125ml) boiling water, or substitute 1 tablespoon lemon juice
2 tablespoons whole coriander seeds
2 teaspoons whole cumin seeds
½ to 1 teaspoon chilli powder
2 ounces (50g) fresh coconut, chopped, or ¼ cup (50ml) unsweetened shredded coconut
3 carrots, peeled and cut in thin strips
¼ pound (100g) green beans, thinly sliced
1 onion, sliced
1 tomato, sliced
1 teaspoon black mustard seeds or ½ teaspoon white
1 teaspoon turmeric

Cook the dal with 1 tablespoon oil and a pinch of salt in 3 cups (750ml) water for 40 to 45 minutes, until the dal is soft but still whole.

Meanwhile, strain the tamarind infusion, discarding the pulp. Crush the coriander, cumin, chilli and coconut together in a bowl or put in a blender and blend until it has formed smooth paste – if using dried coconut, add 2 tablespoons water to prevent the blades sticking. Add the spice mixture to the dal and continue cooking for 20 minutes.

Heat the remaining oil and when very hot add the fresh vegetables, stirring for 3 or 4 minutes. Add the mustard seeds and turmeric and continue to stir-fry for 4 minutes. Add the tamarind water or lemon juice, cover and cook gently for 10 minutes.

Combine the dal mixture and the cooked vegetables and heat through. Serve hot.

Lentilles à la Dijonnaise — *France*

A good dish for people who like French mustard, but it must be the genuine article, not the 'French' mustard made elsewhere which is too vinegary and coarse.

1 cup (250ml) brown lentils
Bouquet garni of 2 sprigs parsley, 1 bay leaf and 1 sprig thyme
1 large onion, thickly sliced or quartered
2 tablespoons butter
¼ cup (50ml) finely shredded ham, or 2 slices bacon, finely diced
1 tablespoon flour
1 cup (250ml) chicken stock
Salt and pepper
½–1 tablespoon Dijon mustard
Garnish:
Chopped parsley

Put the lentils, *bouquet garni* and onion in a saucepan, cover with water and cook for 45 minutes or until tender. Drain, remove the onion and discard the *bouquet*. (The liquid can be used later in a vegetable or lentil soup.)

Melt the butter in a saucepan and lightly fry the onion until soft but not coloured. Stir in the ham or bacon and cook for 3 to 5 minutes. Stir in the flour and, when beginning to colour, stir in the chicken stock and seasoning. Continue to stir as the mixture comes to the boil and cook gently until the sauce thickens (about 5 minutes), stirring occasionally. Add the mustard.

Stir in the drained lentils and serve hot, sprinkled with parsley.

Lentils and Green Onions — *USA*

Serve with roast meat, especially lamb, adding a little of the juices from the roasting pan to the lentils just before serving.

1 shallot or small onion, finely chopped
1 tablespoon butter
1½ cups (375ml) brown or green lentils
1 bay leaf
Salt and pepper
1 cup (250ml) beef or chicken stock
4 green onions, chopped, including the tops green
Garnish:
Chopped parsley

Cook the onion in butter until just golden, then stir in the lentils, bay leaf and salt and pepper. Cover with the stock and bring to the boil, then simmer for 30 minutes or until the lentils are tender and the liquid absorbed.

Stir in the green onions and, if serving with meat, a little hot dripping or gravy. Serve hot sprinkled with parsley.

Curried Soya Beans

1 cup (250ml) soya beans, soaked
2 tablespoons butter
1 small onion, chopped
1 inch (2½cm) fresh ginger, peeled and finely
    chopped
1 teaspoon cumin seeds
½ teaspoon turmeric
1 teaspoon ground cumin
1½ teaspoons ground coriander
1 teaspoon garam masala (p. 68)
¼ teaspoon cayenne pepper
1 teaspoon salt
Garnish:
Chopped parsley

Cook the soya beans for 1½ to 2 hours in plenty of
unsalted water. Drain and reserve the liquid.

Heat the butter in a saucepan and cook the
onion, ginger and cumin seeds for 5 minutes,
stirring, until lightly golden. Stir in the ground
spices and add 2 tablespoons of the bean liquid.
Cook this paste for 2 minutes.

Add the soya beans, salt and enough liquid to
cover. Bring to the boil, cover the pan and simmer
for 30 minutes.

Transfer to a warm dish and sprinkle with
parsley.

Curried Chick-peas   *India*

Serve as the main course of an Indian meal with
rice, fresh chutneys, raitas (yoghurt-based side
dishes), and a spiced vegetable dish or salad.

1 cup (250ml) chick-peas, soaked
6 cloves garlic, crushed
2 onions, chopped
2 tablespoons ghee, butter or oil
1 teaspoon turmeric
1 teaspoon paprika
1 tablespoon ground cumin
1 tablespoon ground coriander
1 tomato, chopped
1 inch (2½cm) piece of cinnamon
Seeds from 3 cardamom pods
2 tablespoons lemon juice
Salt
Garnish:
2 teaspoons garam masala (p. 68)

good, tastes a lot like Italian preparation

Cook the chick-peas with 2 garlic cloves in
unsalted water for 45 minutes to 1 hour. Drain
and reserve the liquid.

In a heavy saucepan cook the onions in the ghee
until deep brown but not too dark. Add 4 tablespoons of the reserved liquid and boil until almost
dry. Stir in the turmeric, paprika, cumin and
coriander.

Add the chick-peas and stir over a low heat for
3 minutes. Add the remaining garlic, tomato,
cinnamon, cardamom seeds and lemon juice.
Cover with the remaining liquid, add salt to taste
and simmer for 30 minutes.

Serve hot, sprinkled with garam masala.

Sautéed Chick-peas   *India*

1 cup (250ml) chick-peas, soaked
2 cloves garlic
1 small dried chilli or $\frac{1}{4}$–$\frac{1}{2}$ teaspoon chilli powder
Ghee or butter
$\frac{1}{2}$ inch (1$\frac{1}{4}$cm) fresh ginger, peeled and chopped
$\frac{1}{2}$ teaspoon turmeric
1 clove garlic, crushed
1 teaspoon garam masala (p. 68)
1 teaspoon salt

Cook the chick-peas with the whole garlic cloves and chilli in lightly salted water for 1 to 1$\frac{1}{2}$ hours. Drain, discarding the liquid, garlic and chilli. Leave for an hour to dry, or dry very thoroughly on paper towels.

Heat about 1 inch (2$\frac{1}{2}$cm) ghee or melted butter in a frying pan, stir in the ginger, turmeric, crushed garlic and chick-peas. Stir, keeping the heat high but making sure the chick-peas don't brown too much. Sprinkle with garam masala and salt and cook for 3 to 5 minutes until golden. Drain and serve.

Split Pea Purée   *Germany*

Good winter food – puréed peas and vegetables baked with fried onions and crisp bacon.

1 pound (450g) split yellow peas
6 ounces (175g) slab bacon
2 carrots, peeled and chopped
1 stick celery, chopped
1 leek, sliced, including half the green top
1 onion, chopped
6 cups (1$\frac{1}{2}$ litres) water
$\frac{1}{4}$ teaspoon dried marjoram
Pepper
1 tablespoon butter or lard
1 large onion, sliced into rings
Salt

Put the peas, bacon, carrots, celery, leek, chopped onion, water, marjoram and a little pepper in a pan. Bring to the boil, partially cover and simmer for 40 to 45 minutes. Drain well.

Remove the bacon, cut into small cubes and cook in the butter until crisp and well browned. Drain on paper towels. Cook the onion rings in the same pan and, when brown, drain on paper towels.

Purée the peas and vegetables, adding salt to taste. Lightly grease an ovenproof dish and transfer the pea purée to this. Cover the purée with the fried onion rings and diced bacon. Bake at 375°F (190°C) for 20 to 25 minutes, until the top is golden. Serve immediately.

Haricots à la Bretonne *France*

Good with most meats, but especially lamb and pork.

1 cup (250ml) Great Northern beans, soaked
1 onion, stuck with 2 cloves
1 carrot, peeled and chopped
Bouquet garni of 1 sprig thyme, 2 sprigs parsley and 1 bay leaf
1 tablespoon unsalted (sweet) butter
2 tomatoes, peeled and chopped
Salt and pepper

Cook the beans in unsalted water with the onion, carrot and *bouquet garni* for 1 hour or until tender. Drain and reserve the liquid, discarding the carrot and *bouquet*.

Remove the cloves from the onion and chop it finely. Heat the butter in a saucepan and cook the onion gently until golden. Add the tomatoes and beans and 1 cup (250ml) of the bean liquid. Season and simmer for 5 minutes. Serve hot.

Porotos Granados *Chile*

This dish is traditionally served with Pebre, a hot chilli vinaigrette-type sauce. It is also good if made with zucchini instead of pumpkin.

1 cup (250ml) Great Northern beans (or saluggia if available), soaked overnight
1 onion, chopped
1 clove garlic, chopped
1 tablespoon olive oil
2 tomatoes, peeled and chopped
2 teaspoons chopped fresh basil, or 1 teaspoon dried
Salt and pepper
½ pound (225g) pumpkin or zucchini, cubed
½ cup (125ml) corn kernels

Cook the beans for 30 minutes only, leaving them in their cooking water.

Cook the onion and garlic in the oil until just golden. Add the tomatoes, basil, salt and pepper and cook gently until the tomatoes form a thick sauce. Stir into the beans and add the pumpkin or zucchini. Cook for 45 minutes or until the beans are tender but not mushy.

Add the corn and continue cooking for a further 10 minutes. Check the seasoning and serve on its own in deep bowls.

Fagioli Piedmontese   *Italy*

A baked bean dish which benefits from the longest, slowest cooking possible. Traditionally it is cooked overnight with raw, soaked beans, but the following is a good variation using dried beans.

1½ cups (375ml) Great Northern beans, soaked overnight
¼ pound (100g) blanched bacon, cut in thin strips
¼ teaspoon ground cinnamon
¼ teaspoon ground mace
3 cloves garlic, chopped
1 shallot or small onion, very finely chopped (minced)
3 tablespoons finely chopped parsley
Salt and pepper

Cook the beans in unsalted water for 1 hour. Drain.

Heat the ovent to 325°F (170°C). Put half the beans in an ovenproof dish and cover with the bacon, spices, garlic, onion, parsley, salt and pepper. Cover with the remaining beans and add enough water to just cover. Cook, covered, for at least 2 hours.

Haricots Tio Lucas   *Spain*

Good with chorizos, pork or bacon.

1¼ cups (300ml) Great Northern beans, soaked
1 bay leaf
1 sprig thyme
2 cloves garlic
Handful of parsley, chopped
1 teaspoon ground allspice
2 tablespoons olive oil
Pepper
Salt

Put all the ingredients except the salt in a saucepan, cover with water and cook for 1 hour.

Remove the thyme. Add salt to taste and reduce the cooking liquid by boiling briskly. Add more oil to taste, check seasoning and serve very hot.

Lobio   *USSR*

Really a warm bean salad, but also a good hot dish for serving with Polish sausage or frankfurters. This can also be served cold.

1 cup (250ml) red kidney beans, soaked
Salt
2 tablespoons oil
1 tablespoon wine or cider vinegar
1 shallot, very finely chopped
Pepper
2 tablespoons finely chopped parsley

Cook the beans in lightly salted water for 45 minutes to 1 hour. Drain well and mix with the other ingredients while still hot.

Judias Blancas   *Spain*

A very easy dish that goes well with meat, eggs or sausages.

1½ cups (300ml) Great Northern beans, soaked
2 tablespoons lard, or 1 tablespoon olive oil if preferred
1 onion, chopped
2 tomatoes, peeled and chopped
1 tablespoon flour
Salt and pepper
½ cup (125ml) dry sherry
1 bay leaf
Garnish:
Chopped parsley

Cook the beans for 1 hour until tender, drain and reserve the liquid.

Heat the lard or oil and cook the onion for 5 minutes. Add the tomatoes and cook 3 minutes. Stir in the flour, salt and pepper, bring to the boil and stir in the sherry. Cook over a high heat for 1 minute. Add the bay leaf, beans and enough cooking liquid to barely cover.

Simmer for 20 minutes and serve hot, sprinkled with parsley.

Fagioli alla Pizzaiola   *Italy*

The quantities given here are for a main course; if you intend serving this with meat, halve the quantities.

1 pound (450g) Great Northern beans (or cannellini if available), soaked
1 tablespoon olive oil
1 onion, chopped
2 cloves garlic, crushed
4 large tomatoes, peeled and chopped
1 tablespoon tomato paste
1 teaspoon oregano
1½ teaspoons chopped fresh basil, or 1 teaspoon dried
1 bay leaf
1 teaspoon sugar
1 teaspoon salt
Pepper

Cook the beans in unsalted water for 1 hour.

Meanwhile, heat the oil in a saucepan and cook the onion and garlic gently for 8 to 10 minutes until soft but not brown. Add the remaining ingredients, bring to the boil and simmer for 45 minutes.

Drain the beans and stir into the vegetables, bring to the boil and serve in deep bowls.

Curried Beans *India*

1½ cups (300ml) Great Northern beans, soaked
2 tablespoons oil
2 tablespoons oil
1 teaspoon whole black mustard seeds
1½ tablespoons ground cumin
1½ tablespoons ground coriander
1 teaspoon grated fresh ginger
½ teaspoon cayenne pepper
½ teaspoon salt
To serve:
Chappatis or other Indian bread and mango
   chutney

Dry the beans. Heat the oil and fry the mustard seeds until they start popping. Add the beans and fry gently for 1 minute. Stir in the spices and salt and cover with water. Bring to the boil, cover and reduce the heat. Cook gently for 1 to 1½ hours, until the beans are tender.

Serve with chappatis and a chutney.

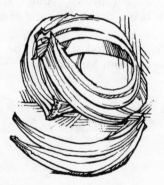

Bean and Tomato Gratinée

1½ cups (375ml) Great Northern beans, soaked
1 onion, finely chopped
1 tablespoon oil
2 cloves garlic, crushed
1 14-ounce (398g) can tomatoes, drained
2 tablespoons tomato paste
Salt and pepper
4 tablespoons butter
⅓ cup (75ml) all-purpose flour
2 cups (500ml) milk
¼ cup (50ml) grated Parmesan cheese
¾ cup (175ml) grated Gruyère cheese
Pinch of ground mace

Cook the beans in unsalted water for 1 hour. Drain.

Cook the onion in the oil until soft. Add the garlic and continue cooking for a few minutes. Stir in the tomatoes, crushing them with the back of a spoon. Add the tomato paste and season. Stir in the beans and transfer to a greased ovenproof dish.

Make a sauce by heating the butter in a saucepan, stirring in the flour and then gradually adding the milk. Bring slowly to the boil, stirring continuously, then simmer for 10 minutes. Stir in the Parmesan and half the Gruyère cheese and season with salt, pepper and mace.

Pour the sauce over the beans, sprinkle with the remaining Gruyère and bake at 400°F (200°C) for 15 to 20 minutes or until the cheese is brown. Serve hot with a green vegetable.

Haricot Bean Mould   *Britain*

This recipe comes from a pre-war cookery book. Serve with a well-seasoned tomato sauce.

1 cup (250ml) Great Northern beans, soaked
1 tablespoon breadcrumbs
2 tablespoons melted butter
1 egg, beaten
Grated nutmeg (to taste)
Salt and pepper
Tomato Sauce:
1½ tablespoons butter
1 tablespoon olive oil
1–2 cloves garlic, crushed
1 small onion, chopped
1 pound (450g) tomatoes, chopped, or 1 14-ounce
    (398g) can tomatoes
1 teaspoon salt
1 teaspoon sugar
1 teaspoon chopped fresh basil or ½ teaspoon dried
1 tablespoon parsley, chopped

Cook the beans for 1 hour. Drain well and purée. Heat the oven to 350°F (180°C). Grease a mould and line with the breadcrumbs, shaking out any excess.

Mix together the bean purée, butter, egg and seasonings. Pour into the mould and stand this in a tray of water. Cover the top of the mould with foil and bake for ¾ to 1 hour or until firm.

Meanwhile, make the sauce: melt the butter and oil and cook the onion very gently for 5 minutes. Add the remaining ingredients and cook over a low heat for 15 to 20 minutes, stirring occasionally. Purée, check the seasoning and reheat to serve.

Turn the mould out on to a hot dish and serve with tomato sauce.

Haricot Beans in Parsley Sauce   *Britain*

Another old recipe, again only as good as the sauce that accompanies it. Parsley sauce is something people either love or hate, but it is particularly good with beans.

1 cup (250ml) Great Northern beans, soaked
1 onion, stuck with 2 cloves
1 tablespoon pure beef dripping or butter
1 tablespoon lemon juice
Parsley Sauce:
2 tablespoons butter
2 tablespoons all-purpose flour
1 cup (250ml) warm milk
Salt and white pepper
Grated nutmeg
2 tablespoons finely chopped parsley
To serve:
Buttered toast

Put the beans, onion and dripping in a saucepan, cover with water and bring to the boil. Cover, reduce the heat and simmer for 1 to 1½ hours.

To make the parsley sauce, melt the butter in a small saucepan and stir in the flour. Add the milk gradually, continuing to stir over a low heat. Cook gently for 8 to 10 minutes and season with salt, pepper and a minute grating of nutmeg. Add the parsley and heat through, stirring, for 1 minute.

Drain the beans and sprinkle with the lemon juice. Pour into individual heated dishes, cover with parsley sauce and serve with hot buttered toast.

Haricot Beans with
Maître d'hotel Butter   *France*

A good dish with roast lamb or pork, chops or ham. The maître d'hotel butter can be made in advance, kept in the refrigerator, and served in small pats or curls as required. This butter is also good with brown or green lentils.

1 cup (250ml) Great Northern beans, soaked
3 ounces (75g) slab bacon
1 small, whole onion
Bouquet garni of 1 sprig thyme, 1 sprig parsley
   and 1 bay leaf
4 tablespoons unsalted (sweet) butter
1 heaped tablespoon finely chopped parsley
2 teaspoons lemon juice

Cook the beans with the bacon, onion and *bouquet garni* in unsalted water for 1 to 1½ hours.
   Remove the bacon, onions and herbs and drain the beans.
   Cream the butter with the parsley and lemon juice and stir into the hot beans.

Pease Pudding   *Britain*

'Pease pudding hot, pease pudding cold, pease pudding in the pot nine days old.' From the days of nursery rhyme, pease pudding has become more a part of folklore than everyday eating as it once was. Traditionally boiled in a pudding cloth and served with boiled beef, it needs a good amount of butter and white pepper.

1 cup (250ml) split yellow or green peas, soaked for
   3 to 6 hours
½ teaspoon salt and white pepper
2 tablespoons butter

Cook the peas in unsalted water for 1 hour or until very soft. Drain and purée them, then return to the pan, season well and heat through carefully.
   Add the butter and serve hot with beef, bacon or pork.

Stir-Fried Bean Sprouts — *China*

Bean sprouts feature in some of the best of Chinese cooking but they must be fresh, not canned. Use mung or soya beans (I prefer the taste of mung sprouts but that is only a personal choice). Spring rolls, chop suey and stir-fried vegetables all include bean sprouts as a major ingredient.

2 tablespoons oil
2 tablespoons chopped green onions
1 pound (450g) bean sprouts
2 tablespoons chicken stock
½ teaspoon sesame oil
½ teaspoon salt
½ teaspoon soya sauce

Heat the oil in a frying pan and stir-fry the green onions and bean sprouts for 1 minute.

Add the stock, sesame oil, salt and soya sauce. Stir for 1 minute and serve.

Steamed Bean Sprouts — *India*

The Chinese do not have the monopoly on bean sprouts, although I have yet to see them served in an Indian restaurant. Very good on their own as part of an Indian meal or to fill crêpes, dosas or omelettes.

½ pound (225g) bean sprouts (mung or urd)
½ tablespoon butter
½ tablespoon chopped parsley
½ tablespoon chopped chives
Salt and pepper
4 teaspoons cold water

Toss the sprouts in the butter in a heavy frying pan. Add the herbs, season and sprinkle with the water. Cover lightly, raise the heat and steam for 3 minutes. Alternatively, steam without the butter in a conventional steamer for 4 minutes and serve with a knob of butter.

Beans Sprouts with Prawns and Pork   *Thailand*

A strong fish sauce, *nam plah*, is usually added – you can use soy sauce to taste or substitute the Chinese *yu chiap* if available. Don't use *blachan* (Indonesian dried shrimp paste) as this is much too strong.

1 tablespoon oil
2 cloves garlic, crushed
3 ounces (75g) lean pork, in very thin slivers
¼ pound (100g) cooked shrimp
Salt and pepper
½ pound (225g) bean sprouts
Pinch of brown sugar
To serve:
Boiled rice

Heat the oil in a heavy pan and stir-fry the garlic and pork for 2 minutes. Add the shrimp and cook for another minute.

Season to taste, stir in the bean sprouts and sugar. Stir for 1½ minutes and serve with rice and a chilli sauce.

Korean Bean Sprouts   *Korea*

We don't hear much about Korean cooking in the West, which is unfortunate as it is quite distinctive, although influenced by the Chinese. The food is spicy – with ginger, garlic and red pepper – and has the same decorative appeal as that from Japan.

1 tablespoon oil, preferably sesame oil
½ pound (225g) bean sprouts
1 to 2 cloves garlic, crushed
½ leek, finely shredded
Pinch of salt
1½ tablespoons soy sauce
1 tablespoon sesame seeds, lightly crushed
¼ teaspoon chopped chilli pepper or ⅛–¼ teaspoon powdered chillis.

Heat the oil and stir-fry the bean sprouts for 1 minute. Add the garlic, leek, salt and soy and cook for ½ minute.

Sprinkle with the sesame seeds and chilli and serve.

Bean Curd and Chinese Cabbage *China*

This could serve several as one of many dishes in a Chinese meal, although it makes a very quick and nutritious meal on its own with rice and soup.

1 tablespoon oil
½ pound (225g) bean curd (p. 141), sliced
½ small Chinese cabbage, shredded
1 green onion, chopped
Salt
2 tablespoons soy sauce

Heat the oil in a wok or large frying pan and stir-fry the bean curd, turning the slices as they brown. Lower the heat slightly, add the cabbage and onion and stir-fry for 1 minute.

Add the salt and soy sauce, stir for ½ minute and serve.

Bean Curd and Mushrooms *China*

An unusual, but good, combination of textures and flavours.

1 tablespoon oil
½ pound (225g) mushrooms, sliced
½ pound (225g) bean curd (p. 141), sliced
½ small onion, finely sliced
2 tablespoons soy sauce
1 tablespoon cornstarch
2–3 tablespoons water

Heat the oil, add the mushrooms and stir-fry for 1 minute. Add the bean curd and stir-fry gently for 2 minutes. Add the onion and continue cooking for 1 minute. Season with soy sauce.

Mix the cornstarch with the water and stir into the vegetables. Cook for a few seconds until the sauce is translucent. Serve hot.

Pasta and Rice Dishes

Pasta e Fagioli   *Italy*

A good winter dish with obvious peasant origins. Serve in deep bowls with bread.

1 cup (250ml) Great Northern beans, soaked
1 ham bone
1 onion, chopped
1 stick celery, chopped
¼ pound (100g) pork rind cut in thin strips
6 ounces (175g) spaghetti
Salt and pepper
To serve:
Grated Parmesan cheese

Put the beans, ham bone, onion, celery and pork rind in a saucepan. Cover with water, bring to the boil and simmer for 1½ hours. Discard the bone. Remove half the beans and purée them.

Bring the remaining beans and liquid back to the boil, add the spaghetti and cook for 10 minutes. Stir in the bean purée and season well. Cook for a further 3 minutes and serve hot with Parmesan cheese sprinkled over.

Lentils and Rice

Serve with cooked vegetables, meat or simply with a tomato and garlic sauce (p. 115).

1 cup (250ml) brown lentils
1 onion, chopped
1 clove garlic, chopped
1 tablespoon olive oil
½ cup (125ml) long-grain rice
2 tomatoes
Vegetable or chicken stock
Salt and pepper

Cook the lentils for approximately 20 minutes and drain.

Cook the onion and garlic gently in the oil. Add the rice and stir until the grains are translucent.

Add the lentils, tomatoes and enough stock to cover. Season and bring to the boil. Lower the heat, partly cover and cook for 20 minutes, until the rice and lentils are tender. Check the seasoning and serve hot.

Pasta and Lentil Bake

This is based on Lasagne al Forno and makes a good vegetarian meal.

1 cup (250ml) brown lentils
1½ tablespoons olive oil
2 cloves garlic, crushed
1 onion, chopped
1 14-ounce (396g) can tomatoes, drained and chopped
1 cup (250ml) stock
2 tablespoons tomato paste
1 bay leaf
1 tablespoon chopped fresh basil, or 2 teaspoons dried
Salt and pepper
8–10 ounces (225–275g) lasagne
4 tablespoons butter
⅓ cup (75ml) all-purpose flour
2 cups (250ml) milk
¾ cup (175ml) grated Parmesan cheese

Cook the lentils for 30 to 35 minutes. Drain. Pre-heat the oven to 425°F (220°C).

Heat the oil in a saucepan and add the garlic and onion, cooking them gently until soft. Add the lentils, tomatoes, stock, tomato paste, bay leaf, basil, salt and pepper. Stir well and cook gently while you prepare the lasagne.

Cook the pasta according to the packet directions and drain. Meanwhile, make a sauce by heating the butter in a saucepan, stirring in the flour and then gradually adding the milk, stirring continuously. Simmer for 10 minutes, season the sauce and add ½ cup (250ml) of the cheese.

Arrange the lentil sauce, pasta and cheese sauce in layers in an oven-proof dish, starting with a layer of lentil sauce and finishing with a layer of cheese sauce. Sprinkle the top with the remaining cheese and bake for 20 minutes until the top is bubbling and golden brown. Serve hot.

Riz et Pois — *Haiti/Jamaica*

Variations of riz et pois (the pois being red kidney beans, not peas) occur in most Caribbean islands. Sometimes the recipe includes meat, coconut and vegetables or, as in this case, it just makes a simple risotto-type dish which is very good with pork dishes.

½cup (125ml) red kidney beans, soaked
Salt
2 tablespoons lard
1 cup (250ml) long-grain rice
Pepper

Cook the beans in lightly salted water for 1 hour. Drain, reserving the liquid, and make it up to 2 cups (500ml) with cold water.

Heat 1 tablespoon lard in a heavy saucepan until very hot, but not smoking, and stir in the rice. Cook for 1 minute, lower the heat and pour in the bean liquid, season and simmer for 20 minutes or until the rice is cooked and dry. Remove from the heat and cover with a clean teatowel and a lid.

Heat the remaining lard in a frying pan, stir in the beans and cook for 1 or 2 minutes without browning them. Serve the rice on a heated dish, with the beans on top.

Middle Eastern Rice and Lentils — *Middle East*

Good enough to eat alone, with a bowl of fresh yoghurt.

1 cup (250ml) brown lentils
1 large mild onion
2 tablespoons olive oil or butter
1 cup (250ml) long-grain rice
Salt and pepper

Cook the lentils in lightly salted water for 30 minutes. Drain, reserving the liquid. Cut the onion in half. Slice one half and finely chop the other.

Heat 1 tablespoon of oil in a saucepan and fry the chopped onion. Add the rice and stir until translucent.

Add the lentils and enough of the lentil liquid to cover the rice. Season, cover and simmer for 20 minutes until the rice is tender and dry.

Fry the sliced onion in the remaining oil until brown. Serve the rice and lentils hot, topped with the fried onion.

Moros y Cristianos *Cuba*

Moors and Christians, so called because of the use of black beans and white rice, is a good dish on its own or to serve with meat. Black beans are especially popular in Cuba where, as in most Caribbean islands, beans in general are a major source of nutrition.

1 cup (250ml) black beans, soaked
2 tablespoons oil
2 slices blanched bacon, coarsely chopped
1 small onion, chopped
2 cloves garlic, crushed
½ small green pepper, seeded and finely chopped
1¼ cup (300ml) long-grain rice
Salt and pepper

Cook the beans in unsalted water for 1 to 1½ hours or until soft. Drain. Mash 3 tablespoons of the beans with a fork.

Heat the oil in a heavy pan and cook the bacon until crisp. Remove the bacon and drain on paper towels.

Add the onion, garlic and green pepper to the oil and cook gently for 5 minutes.

Add the bean paste and stir well, then add the rice, beans and bacon. Season and add enough water to cover the rice. Bring to the boil, cover and simmer for 20 minutes.

Fluff the rice up with a fork and serve hot.

Hoppin' John *USA*

Real 'soul' food from the Southern states, usually served with pig's cheek, hocks, bacon and collards (similar to kale). Very good on its own or with pork, ham or sausages.

1 cup (250ml) black eyed peas, soaked
1 teaspoon salt
1 cup (250ml) long-grain rice
1 onion, chopped
1 tablespoon oil
4 tomatoes, peeled and chopped
½ teaspoon cayenne pepper
Black pepper

Cook the peas in lightly salted water for 40 minutes. Add the rice and continue to simmer.

Meanwhile, cook the onion in the oil until soft but not brown. Stir in the tomatoes, salt, cayenne and pepper and add this to the rice and peas. Cover and simmer on the lowest heat for 15 minutes, until the liquid is absorbed and the peas tender but still whole. Serve hot.

Any leftovers are good served as a salad with oil and lemon, some raw onion rings and chopped parsley.

Pasta e Ceci   *Italy*

A very simple dish but, with good olive oil and
fresh Parmesan, it makes an excellent and filling
lunch. Any type of pasta can be used, but the
more decorative shapes look better.

1 cup (250ml) chick-peas, soaked
Salt
6 ounces (175g) pasta
2 tablespoons olive oil
¼ cup (50ml) freshly grated Parmesan cheese

Cook the chick-peas in lightly salted water for 1
to 1½ hours or until tender. About 20 minutes
before they are ready, cook the pasta separately
according to the instructions on the packet.

   Drain both the pasta and the chick-peas, combine in a warm bowl and toss them in the oil and
cheese.

Red Beans and Rice   *Jamaica*

This goes well with sautéed or grilled chicken.

1 cup (250ml) red kidney beans, soaked
1 onion, finely chopped
1 small green pepper, seeded and diced
1 stick celery, diced
1 cup (250ml) coconut milk (p. 81)
Salt and pepper
¼ teaspoon dried thyme
1 cup (250ml) long-grain rice

Cook the beans for 45 minutes. Drain and reserve
the liquid.

   Place the onion, pepper, celery and coconut milk
in a saucepan. Add ½ cup (125ml) of the bean
cooking liquid, bring to the boil, cover and simmer
for 20 minutes.

   Season with salt, pepper and thyme and add the
beans. Bring to the boil and add the rice, lower
the heat and simmer, partly covered, for 15 to 20
minutes until the rice is dry and fluffy. Serve
immediately.

Croquettes and Burgers

Lentil Croquettes (1)   *Poland*

These are very popular with children. The lentil purée must be thick and dry. Fry in oil if they are to be served alone as a vegetarian lunch, or cook in a suitable fat if accompanying a meat dish. Very good with bacon or ham dishes and both English and Polish sausages.

1½ cups (375ml) brown or green lentils
Salt
1 egg, beaten
Pepper
Fresh or dried breadcrumbs
Oil, lard, dripping or bacon fat

Cook the lentils for 45 minutes to 1 hour in lightly salted water. Drain well and purée.

Add half the egg to the purée, together with salt and pepper to taste. Shape into flat round cakes and coat evenly with the remaining egg and the breadcrumbs.

Shallow fry for 4 minutes on each side until golden.

Lentil Croquettes (2)   *Britain*

Good with eggs, sausages, bacon, or alone with a colourful vegetable such as grilled tomatoes or creamed carrots.

1 cup (250ml) red lentils
½ cup (125ml) cooked rice
1 egg, separated
Salt and pepper
1 tablespoon oil or butter
1 small onion, finely chopped
Flour
Breadcrumbs
Oil (for deep frying)

Cook the lentils in unsalted water for 45 minutes until soft. Drain well and mash. Add the rice, egg yolk, salt and pepper.

Heat the oil and cook the onion until soft and lightly browned. Add to the lentils. Leave until cold and form into croquettes or flat cakes, flouring your hands before shaping them.

Lightly beat the egg white and egg and crumb the croquettes. Heat the oil to 375°F (190°C) and fry until crisp and golden. Drain well and serve hot.

Soya Bean Croquettes

I suppose beanburgers would be a more appropriate name. These croquettes or cakes can be fried or baked, and are as good cold with a salad as they are hot with vegetables or bread.

1 cup (250ml) soya beans, soaked
1 onion, minced or grated
1 carrot, peeled and grated
1 clove garlic, finely chopped
⅔ cup (150ml) peanuts
3 tablespoons sesame seeds
2 tablespoons tahina paste
1 egg, beaten
3 tablespoons breadcrumbs, preferably made from
    whole wheat bread
Salt and pepper
½ teaspoon mixed dried herbs (celery salt is particularly good)
1 teaspoon Kitchen Bouquet (optional)
Oil

Cook the beans for 2 hours. Drain and purée.

Grind the peanuts to a fine meal in a blender or mortar, but don't let them become a paste.

Mix together all the ingredients except the oil so that a stiff paste is formed. Shape into cakes or croquettes.

Heat the oil and fry the bacon on both sides, or bake at 350°F (180°C) for 25 minutes.

Chick-pea Croquettes

Serve these hot or cold with a vegetable dish or salad. This mixture can also be used as a sandwich filling when cold, in which case it should be mixed with a little fruit chutney.

If bessan flour is not available, breadcrumbs can be substituted in this particular recipe.

⅔ cup (150ml) chick-peas, soaked
1 onion, finely chopped
¼ red pepper, finely chopped
2 tablespoons finely chopped parsley
Juice of 1 lemon
Salt and pepper
1 tablespoon soy sauce
1 tablespoon bessan flour (p. 142)
Oil

Cook the chick-peas for 1 hour, drain and mash until smooth.

Stir in the onion, pepper and parsley together with the lemon juice, seasoning and soy sauce. Finally add the bessan flour, stirring well. Form the mixture into flat 'burger' shapes or into cylindrical croquettes.

Heat the oil and fry the chick-pea mixture for approximately 5 minutes on each side until crisp and golden. Drain well before serving.

Spiced Soya Cutlets

Similar to chick-pea croquettes but with a coarser texture. If preferred, the cutlets can be coated with breadcrumbs before cooking. The garam masala can be replaced with a mixture of ground spices to taste – cumin and coriander with a little white pepper and turmeric are a good combination, especially if the cumin and coriander are roasted lightly before grinding.

½ cup (125ml) soya beans
1 onion, finely chopped
⅓ cup (75ml) unsalted peanuts, chopped
⅓ cup (75ml) cashew nuts, chopped
2 tomatoes, peeled and finely chopped
2 tablespoons lemon juice
Salt
¼ teaspoon cayenne pepper
1 tablespoon (or more to taste) garam masala
  (p. 68)
Oil

Cook the beans for 2 hours, drain and mash.

Add the onion and the nuts, tomatoes, lemon juice, salt, cayenne and garam masala. Shape into small flat cakes and fry for approximately 5 minutes each side in hot oil, regulating the heat to avoid burning.

Drain and serve hot with a rice salad or a raita (yoghurt with chopped cucumber, banana, or coconut, and salt and cayenne or cumin to taste).

Zirus Pikas   *Latvia*

Potato and split pea croquettes which can be served with many meat dishes, particularly sausages, ham or bacon. They are also good shaped into flat round cakes and fried in bacon fat.

1 cup (250ml) split green or yellow peas, soaked
Salt
1 large or 2 medium potatoes, cooked and mashed
¼ pound (100g) bacon, finely chopped or minced
1 tablespoon oil
Pepper
¼ teaspoon ground allspice

Cook the peas in lightly salted water for 45 minutes. Drain and purée until very smooth. Mix with the potato.

Lightly fry the bacon in the oil, then add all the remaining ingredients with the bacon to the potato mixture. Mix well, add salt to taste and shape into balls – an icecream scoop is ideal for this. Serve immediately.

Aduki Beancakes

1¼ cups (300ml) aduki beans, soaked
Salt
½ cup (125ml) wheatgerm
Oil
Soy or Tamari sauce

Cook the beans for 2 hours. Drain well and mash.
Mix with the salt and wheatgerm. Shape into
small cakes, about 3 inches (6cm) across.

Roll in a little extra wheatgerm and shallow-
fry in the oil until crisp and golden.

Sprinkle with the sauce and turn the cakes.
When both sides are crisp, drain well on paper
towels and serve hot.

Lentil Cakes

1 cup (250ml) lentils
3 potatoes, peeled, cooked and mashed
1 tablespoon finely chopped parsley
1 tablespoon lemon juice
2 tablespoons melted butter
Salt and pepper
Flour
Bacon fat

Cook the lentils for 30 minutes, drain well and
mash. Add to the potato, making sure they are
thoroughly blended.

Stir in the parsley, lemon juice and butter and
season well. Form into flat round cakes and
lightly flour them on both sides. Cook until brown
in very hot bacon fat and serve with bacon.

Casseroles and Stews

Soya and Corn Casserole

Serve either as a side dish with meat or as a
separate course on its own.

½ cup (125ml) soya beans, soaked
¾ cup (200ml) corn kernels
1 clove garlic, crushed
1 onion, finely chopped
4 tomatoes, chopped
2 tablespoons tomato paste
Salt and pepper
Pinch of paprika or cayenne pepper
Vegetable stock
Butter

Cook the beans for at least 1 hour. Drain.

Combine all the ingredients except for the stock
and butter and add enough stock to just cover.
Season well, cover and bake at 400°F (200°C) for 25
to 30 minutes. Serve with a lump of butter on top.

Khichiri (1)   *India*

There are so many versions of this rice and dal stew, that I have included more than one recipe. Ingredients vary from region to region but the rather soupy consistency remains.

2 tablespoons ghee, butter or oil
⅔ cup (150ml) rice
½ (125ml) masur dal or red lentils
1 teaspoon salt
½ inch (1¼cm) fresh ginger, peeled and finely chopped
3 cloves
9 peppercorns
Seeds from 3 cardamom pods
3 bay leaves
1 inch (2½cm) piece of cinnamon
3 onions, sliced

Melt half the ghee and toss the rice and dal in it until lightly coated.

Stir in the salt, ginger, cloves, peppercorns, cardamom seeds, bay leaves and cinnamon and cover with water. Bring to the boil and simmer for 25 to 30 minutes, adding more water if it appears dry.

Fry the onions in the remaining ghee and sprinkle over the khichiri before serving. Serve hot.

Kitchedi (Khichiri 2)   *India*

1 onion, thinly sliced
2 tablespoons ghee, butter or oil
1½ tablespoons whole cumin
½ teaspoon whole allspice
6 peppercorns
Seeds from 6 cardamom pods
1 inch (2½ cm) fresh ginger, peeled, finely chopped and crushed
½ teaspoon turmeric
⅔ cup (150ml) long-grain rice, preferably basmati
½ cup (125ml) masur dal or red lentils
2 teaspoons salt
3 tablespoons tamarind water (made by infusing 1 tablespoon tamarind pulp in 3 tablespoons warm water and straining) or lemon juice.

To serve:
Chutney

Cook the onion gently in the ghee or butter for about 8 minutes.

Meanwhile, pound the whole spices with the ginger and turmeric, using a mortar and pestle or a blender. Stir the spice paste into the onions, add the rice and dal and stir well. Sprinkle with salt, tamarind water and enough cold water to cover. Bring to the boil, cover and simmer for 30 minutes.

Cover with a clean teatowel and the lid and leave for 5 minutes off the heat. Fluff up the rice and serve hot with a fresh chutney.

Red Bean Casserole

2 slices bacon, chopped
2 tablespoons oil
1 onion, chopped
3 cloves garlic, crushed
1 carrot, peeled and chopped
1 red pepper, seeded and sliced
1 cup (250ml) red kidney beans, soaked
3 tomatoes, peeled and chopped
1 tablespoon tomato paste
1 bay leaf
1 sprig parsley, 1 sprig thyme and 1 sprig marjoram, tied together
Salt and pepper
2 cups (500ml) vegetable or chicken stock
¼ pound (100g) chorizo or garlic sausage (optional)

Cook the bacon in the oil until brown. Remove the bacon and cook the onion, garlic, carrot and pepper for 5 minutes until browned.

Add the beans, tomatoes, tomato paste, herbs, seasoning and stock. Bring to a boil, then transfer to a casserole. Add the bacon and sliced chorizo.

Bake in a covered casserole at 350°F (180°C), for 1 to 1½ hours. Check seasoning and serve piping hot with rice, or in deep bowls with bread.

Eggplant and Chick-Pea Stew

1¼ cups (300ml) chick-peas, soaked
2 tablespoons olive oil
1 onion, chopped
2 cloves garlic, chopped
1 large or 2 small eggplants, cubed
¾ pound (350g) tomatoes, peeled and chopped
½ teaspoon ground cumin
Pinch of cayenne pepper
Salt and pepper

Cook the chick-peas in unsalted water for 1 hour until just tender. Drain.

Heat the oil in a saucepan and cook the onion and garlic gently, without browning, for 5 minutes. Add the eggplant and cook gently for 5 minutes. Add the tomatoes, spices, salt and pepper and stir well.

Heat through, add the chick-peas and, when hot, serve with fresh bread.

This is also good served cold with a little lemon juice.

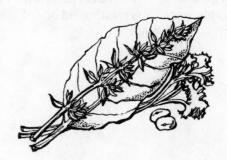

Soya Bean Casserole

Soya beans seem to benefit from other strong flavours. The herbs can be varied to suit your own taste.

1 cup (250ml) soya beans, soaked
2 tablespoons oil
1 large mild onion, chopped
2 cloves garlic, crushed
2 sticks celery, chopped
3 tomatoes, peeled and chopped
2 tablespoons chopped parsley
2 to 3 tablespoons oil
2 to 3 tablespoons chopped fresh herbs (chives, lemon thyme, celery leaves, basil, oregano, etc.)
Salt and pepper

Cook the beans in unsalted water for 1½ to 2 hours. Drain.

Preheat the oven to 350°F (180°C). Heat the oil and cook the onions, garlic and vegetables for about 8 minutes until soft but not browned. Stir in the herbs and seasonings.

Put the beans in an ovenproof dish and stir in the vegetable mixture. Barely cover with water or vegetable stock, cover with a lid or foil and bake for 1 hour. Serve hot.

Haricot Giurvech *Yugoslavia*

Giurvech, guivetch (and other spellings) cover a wide range of Balkan vegetable stews which sometimes include meat. This recipe is for a simple bean and vegetable stew, baked in a thickened sauce in the oven.

1 cup (250ml) Great Northern beans, soaked
2 tablespoons oil
1 onion, chopped
1 green or red pepper, seeded and diced
3 tomatoes, peeled and chopped
½ teaspoon salt
1 teaspoon paprika
1½ tablespoons all-purpose flour
To serve:
Crusty bread

Cook the beans, with 1 tablespoon of the oil, in enough water to cover for 1 hour. Drain and reserve 1 cup (250ml) of the liquid.

Heat the remaining oil and brown the onion. Lower the heat, add the pepper and cook gently for 5 minutes. Add the tomatoes, season and cook very gently for 10 minutes.

Meanwhile, heat the oven to 300°F (150°C) and lightly oil an ovenproof serving dish or casserole. Stir the beans into the cooked vegetables and transfer to the dish.

Stir the flour into the pan in which the vegetables were cooked. Gradually stir in the bean liquid, bring to the boil – still stirring – and pour over the giurvech. Bake, covered, for 30 minutes. Uncover and bake for a further 30 minutes. Serve hot with fresh crusty bread.

Spicy Red Bean Stew   *Mexico*

1 cup (250ml) red kidney beans, soaked
2 cloves garlic, crushed
1 onion chopped
1 tablespoon oil
1 teaspoon ground cumin
½–1 teaspoon chilli powder
⅔ cup (150ml) beef stock
1½ tablespoons chopped parsley
1 bay leaf
½ teaspoon oregano
Salt and pepper
1 tablespoon tomato paste

Cook the beans, garlic and onion together in unsalted water for 45 minutes to 1 hour, until tender but still whole. Drain.

Heat the oil in a heavy saucepan. Remove from the heat and stir in the spices to make a warm paste. Stir in the beans, stock, parsley, bay leaf and oregano. Season well and return to the heat. Simmer for 30 to 40 minutes.

Stir in the tomato paste, check the seasoning and serve hot, either alone or with rice or tortillas.

Spinach and Lentil Stew   *Persia*

A delicious dish if you like spinach.

1 cup (250ml) brown lentils, soaked for 1 hour
¾ pound (350g) spinach, thoroughly washed
1 teaspoon ground coriander
½ teaspoon ground cumin
2 cloves garlic, finely chopped
Salt and pepper
To serve:
Butter

Cook the lentils in just enough water to cover for 40 minutes, or until tender.

Gently cook the spinach in a covered pan (do not add any water), or steam it, for 5 or 6 minutes. Drain any liquid from the spinach and chop finely.

Drain the lentils and stir in the spices, garlic and spinach. Season with salt and pepper and serve hot with a big knob of butter on top.

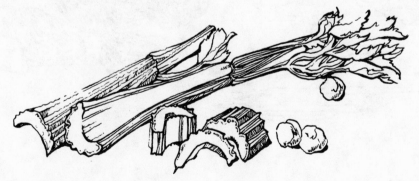

Bean and Nut Stew

A very easy dish which can be varied by changing the nuts, beans and herbs used. Equally good served over brown rice or served cold with plenty of fresh lemon juice.

½ cup (125ml) soya beans, soaked
1 cup (250ml) vegetable stock
½ cup (125ml) Great Northern beans, soaked
½ cup (125ml) red kidney beans, soaked
1 onion, finely chopped
1 tablespoon oil
⅔ cup (150ml) walnuts or pecans, roughly chopped
⅔ cup (150ml) peanuts, roughly chopped
Salt and pepper
1 tablespoon finely chopped fresh basil
1 tablespoon finely chopped parsley

Cook the soya beans in the stock for 1 hour. Add the Great Northern and kidney beans and continue to cook for another hour, adding a little water if it gets too dry. Drain, reserving the liquid.

Cook the onion gently in the oil until soft but not brown.

Add the nuts, stir for a few minutes and add the beans with enough of their liquid to moisten. Heat through. Check the seasoning and stir in the basil and parsley.

Fruit and Sweetened Dishes

Baked Pinto Beans with Honey

Good with pork dishes or served alone with a green salad.

1½ cups (375ml) pinto beans, soaked for 3 to 4 hours
1 onion, grated
4 tablespoons honey
2 cloves garlic, crushed
Salt and pepper
¼ teaspoon paprika

Cook the beans for 40 minutes. Drain. Preheat the oven to 400°F (200°C).

Stir the onion into the beans together with the honey and garlic. Season well with salt, pepper and paprika and transfer to a casserole.

Cover and bake for 30 minutes, adding a little water after 20 minutes if it has become too dry.

Boston Baked Beans *USA*

Baked beans have changed, somewhat for the worse, since the pioneering days when dishes of beans flavoured with mustard, molasses and salt pork were baked with batches of bread.

Alternative flavourings include 1 tablespoon tomato paste and/or 1 tablespoon rum, both of which are added with the molasses. Serve the beans on their own with fresh bread or with frankfurters and sliced onions.

1 pound (450g) small white or Great Northern
   beans, soaked
2 onions, each stuck with 2 cloves
½ pound (225g) salt pork, cubed
½ cup (125ml) dark brown sugar
2 tablespoons molasses
2 teaspoons dry mustard
1 teaspoon salt
Pepper

Cook the beans in unsalted water for 1 hour. Drain and reserve the liquid.

Use a heavy casserole with a lid and place the onions at the bottom. Score the salt pork cubes and put on top of the onions. Cover with the beans.

Mix the sugar, molasses, mustard, salt and a good amount of black pepper with 1 cup (250ml) of the bean liquid. Pour over the beans, stir gently and cover the casserole. Bake for at least 3 hours at 250°F (125°C) adding more liquid if too dry. If the dish is too liquid, remove the lid for the last 30 minutes of cooking.

Swedish Brown Beans
(Bruna Bönar) *Sweden*

A sweet and sour dish in which the cooking liquids are reduced to a thick sweet sauce. Traditionally served with boiled bacon, sausages, or slices of salt pork coated in egg and breadcrumbs and fried.

½ pound brown or small white beans, soaked in 3
  cups (750ml) water overnight
⅓ cup (75ml) white vinegar
⅓ cup (75ml) corn syrup
1 tablespoon dark brown sugar
1 teaspoon salt

Cook the beans in their soaking water for 45 minutes.

Stir in the other ingredients and cook very gently for 1 hour. Keep checking that it doesn't boil dry, and top up with a little more water if necessary. Serve very hot.

Honeyed Lima Beans

This recipe is in the tradition of Bruna Bönar and Boston Baked Beans, where the use of a sugar, in this case honey, provides a glazed sauce to the finished dish.

1 cup (250ml) lima beans, soaked
2 tablespoons oil
1 large onion, sliced
½ teaspoon mustard
3 tablespoons honey
Salt and pepper

Cook the lima beans in unsalted water until tender, 40 minutes to 1 hour. Drain and reserve the liquid. Heat the oven to 350°F (180°C).

Heat the oil and cook the onion for 8 or 10 minutes until just beginning to brown. Stir in the mustard and honey, salt to taste and plenty of pepper. Add the beans and enough of their cooking liquid to barely cover.

Transfer to a casserole and bake, uncovered, for 20 minutes, then cover and cook for 40 minutes. Serve hot.

Tomato Baked Beans   *USA*

Although this may sound rather like the canned variety the similarity ends there. Good with meat, sausages or on toast.

1 cup (250ml) Great Northern beans, soaked
3 tablespoons tomato paste
1 onion, finely chopped
2 teaspoons brown sugar
2 teaspoons dry mustard
Salt and pepper

Cook the beans in unsalted water for 40 minutes. Heat the oven to 325°F (170°C).

Drain the beans, reserving the liquid. Stir in the other ingredients, adding enough of the cooking liquid to moisten.

Pour into a casserole and cook, covered, for 1½ to 2 hours. Check that it does not boil dry, adding more liquid if necessary.

Bean and Apple Casserole

Good with bacon or sausages.

1 cup (250ml) Great Northern or butter beans, soaked
2 firm eating apples, peeled, cored and quartered
½ tablespoon brown sugar
½ teaspoon cinnamon
1 tablespoon red wine vinegar
2 tablespoons butter or lard
Salt

Cook the beans in unsalted water until tender, about 1 to 1½ hours. Drain, reserving the liquid. Preheat the oven to 350°F (180°C).

Stir the apples, sugar, cinnamon, vinegar, butter and salt into the beans. Transfer to an oven-proof dish and just cover with some of the reserved liquid. Cover with a lid or foil and bake for 35 to 40 minutes.

Westfälisches Blindhun — *Germany*

This Westphalian dish is perfect winter food, served in deep bowls with fresh bread.

1 cup (250ml) Great Northern beans, soaked
½ pound (225g) thickly sliced bacon
3 cups (750ml) water
1 pound (450g) firm cooking apples, peeled, cored and thickly sliced
1 pound (450g) firm pears, peeled, cored and thickly sliced
¾ pound (350g) green beans, sliced
2 carrots, peeled and diced
3 potatoes, peeled and diced
Salt and pepper

Put the beans, bacon and water in a large pan. Cook for 1 hour or until the beans are tender.

Add the fruit, vegetables, salt and pepper and simmer gently for 25 to 30 minutes. Check the seasoning and serve very hot.

Beans and Prunes — *Germany*

In Germany dried fruits are often called upon to add variety and flavor to winter dishes, as is demonstrated by their Schlesisches Himmelreich (pork and dried fruit casserole with dumplings), Beerenpie (dried fruit and bacon tart), etc. This simple recipe is especially good with a rich meat like goose or pork.

⅔ cup (150ml) Great Northern beans, soaked
1 cup (250ml) prunes, halved, stoned and soaked overnight
Salt and pepper
Pinch of sugar

Cook the beans and prunes together in the prune soaking liquid for 1½ hours. Drain, add salt, pepper and sugar and serve hot.

Home-made Bean Curd (Tofu)

Begin 48 hours in advance. The most important factor in the preparation of bean curd at home is the temperature at which it ferments. A warm place such as an oven with a pilot light is good; it is important that the temperature remain constant. If you are fortunate enough to have a Chinese supermarket in the neighbourhood, try some commercially made fresh curd, so that you know the texture to aim for.

The quantities given produce a soft-textured curd, rather like a very soft cheese, and the curd needs to be drained before being added to the chosen recipe. Although bean curd is often added to salads as it is, I find the flavour too bland on its own. It has an obvious affinity with soy sauce and can be added to most Chinese vegetable dishes.

1 cup (250ml) soya beans
¼ cup (50ml) fresh lemon juice

Soak the beans for 24 hours, changing the water several times. Drain. Grind the beans to a pulp in a mortar or a blender. Weigh the bean pulp and place in a large pan. Add twice the weight of the beans in cold water, stir and bring to the boil. Lower the heat and simmer for 1 hour, or cook for 20 minutes in a pressure cooker. Watch that the bottom does not burn – the mixture at this stage will resemble a fairly thick soup.

Allow the liquid to cool slightly and strain through muslin over a non-metallic bowl, squeezing to extract all the liquid. You should have a quantity of cream-coloured liquid resembling a slightly gelatinous cream.

Discard the pulp and stir the lemon juice into the liquid. Cover with a damp cloth and leave in a warm place for up to 12 hours. Given an even temperature, the curd can set in as little as 4 hours. It should be firm to the touch and hold its shape when cut.

Drain on muslin to remove any water and store in the refrigerator until needed. Although it will keep for 2 to 3 days without harm, it does not keep as well as the commercial variety.

You can serve bean curd in a number of ways: add it, diced, to soups and vegetable dishes; use it in recipes from this book (pp. 119 and 142); grill it, or add it to sandwiches. But always remember that bean curd needs other ingredients with strong flavours to compensate for its blandness, and take this into account when using it for cooking.

Miso/Tofu Spread   *Japan*

This spread does not need seasoning as the miso is salty enough on its own. Serve on whole wheat bread with a salad, or on toast.

2 ounces (50g) tofu (bean curd)
1 tablespoon lemon juice
3 tablespoons miso, puréed
1 tablespoon grated orange peel

Mash the tofu with the lemon juice. Stir in the miso and orange peel, adding a few drops of water to achieve a spreading consistency.

Tofu and Noodles   *Japan*

10–12 ounces (275–350g) noodles, preferably
   Japanese udon
3 cups (750ml) chicken stock
½ pound (225g) bean curd (p. 141), sliced
Oil
1½ tablespoons soy sauce
2 tablespoons dry sherry
Pinch of salt
8 large green onions, chopped

Cook the noodles in the stock until soft. Meanwhile, fry the bean curd in the oil until both sides are well browned.

Combine the soy and the sherry in a small pan, add 1 or 2 tablespoons of stock and bring to the boil. Add a little salt and stir the fried bean curd carefully in the sauce for a few minutes.

Pour the noodles and the stock into individual soup bowls and divide the bean curd amongst them. Sprinkle with the green onions.

Flours and Batters

In India and the Middle East lentils and chick-peas are ground to a flour which is used mostly for frying batters. The number of Indian fried 'snacks' alone would fill a book, so I have just included a selection of some of the more unusual ones.

Bessan   *India*

This is the chick-pea flour which is available from Indian grocers. To make your own you need a blender or food processor and a lot of patience! Use split chick-peas or channa dal as they are softer.

Heat a heavy pan, ungreased, and toast the dal lightly – but not enough to colour it. Remove from the pan and grind in small quantities at full speed, then sieve. Regrind any large particles until you have a fine, yellow flour. Store in an airtight container.

Bessan Frying Batter

India

A crusty, reddish-gold batter which should be used to coat chicken pieces or for vegetable fritters – superb wth sliced eggplant or zucchini.

1 cup (250ml) bessan (p. 142)
Pinch of pepper
1 tablespoon ground coriander
Seeds from 1 cardamom pod, ground or crushed
Pinch of cayenne pepper
1 teaspoon salt
4 tablespoons plain yoghurt

Mix together the flour, spices and salt and beat in the yoghurt. Leave for 45 minutes before use.

Frying Batter for Fish

India

Use to coat cubes or strips of white fish, then deep fry in oil.

⅓ cup (75ml) bessan (p. 142)
2 tablespoons rice flour
½ teaspoon ground cumin, preferably roasted and freshly ground
Pinch of salt
Pinch of cayenne pepper
3 tablespoons water

Mix the dry ingredients together and beat in enough water to make a smooth, thick batter. Leave for 30 minutes before use.

Onion Pakoras (Bhajias)

India

Serve as a snack or as part of an Indian vegetarian meal. You can replace the chopped onions with small whole cauliflower florets, cubed cooked potato, diced cucumber, etc.

½ cup (125ml) bessan (p. 142)
½ teaspoon turmeric
1 teaspoon ground coriander
1 teaspoon ground cumin
½ teaspoon paprika
1 teaspoon garam masala (p. 68)
1 teaspoon salt
4 tablespoons water
1 large onion, chopped
1 teaspoon chopped parsley
Oil

Mix together the bessan, spices and salt. Beat in the water slowly until a smooth, thick batter is formed. Leave for 30 minutes, then stir in the onion and parsley.

Heat the oil to 375°F (190°C) and drop in 1 tablespoonful of the batter mixture at a time. Fry until a deep golden brown, turn and brown the other side (about 3 minutes for each side). Drain well and keep warm while you fry the rest of the batter.

Prawn (Shrimp) Pakoras: Prepare the batter as above but limit the flavouring to a pinch of salt, pepper or cayenne and a pinch of turmeric. Dip cooked shelled shrimp into the batter and deep fry for 3 minutes.

Pakora Raita   *India*

Crisp bessan batter balls in spiced yoghurt – an excellent side dish for a vegetarian Indian meal. Very popular with children, as are all these deep fried bessan dishes — and more nutritious than potato chips!

¼ cup (50ml) bessan (p. 142)
Salt
3 tablespoons water
Oil (for deep frying)
1 cup (250ml) plain yoghurt, chilled
½ teaspoon ground cumin
Pinch of cayenne pepper
Garnish:
1 tablespoon chopped parsley

Mix the bessan, a pinch of salt and the water to a smooth pouring batter.

Heat 2 inches (5cm) oil to about 350°F (180°C). Using a large-holed colander, a perforated spoon or a grater, pour the batter through the holes into the oil. Fry the little balls for a few seconds and as they brown, remove and drain on kitchen paper. Continue until all the batter has been used.

Mix together the yoghurt, a pinch of salt and the spices. Gently fold in the pakoras and reserve a few for decoration. Sprinkle them on the top of the yoghurt together with the parsley.

Chiura

A delicious combination of chick-pea flour noodles, peanuts and spices. If you cannot get hold of rice flakes, the wheat or barley flakes which are obtainable in health food shops are a good substitute. This is very good nibbling food for parties, picnics, etc.

2 tablespoons bessan (p. 142)
1 teaspoon salt
3 tablespoons water
Oil or ghee (for deep frying)
½ cup (125ml) shelled unsalted peanuts
⅓ cup (75ml) phoa (rice flakes), or stone-ground wheat or barley flakes
Pinch of turmeric
1 teaspoon sugar
Pinch of cayenne pepper

Mix the bessan, half the salt and the water to a thin pouring batter. Leave for 30 minutes.

Pour approximately 2 inches (5cm) oil or ghee into a pan and heat to 350°F (180°C). Use a large-holed colander and pour the batter through the holes, moving the colander so that small noodles are formed. Fry for 1 minute, remove with a perforated spoon and drain. Continue until all the batter is used.

Reheat the oil and fry the peanuts for 2 minutes. Remove and drain. Pour off most of the oil, leaving about 1 tablespoon. Stir in the phoa or wheat flakes and the turmeric and fry for 1 minute or until golden. Remove and drain.

Combine the noodles, peanuts, flakes, sugar, cayenne and remaining salt in a bowl, turning to mix thoroughly.

Store in an airtight container until ready to use.

Pancakes and Fritters

Dosas

Dosas, or doshas, are thin, spiced pancakes which are either fried or steamed and served either as a bread or stuffed with a savoury vegetable filling. Begin the day before you intend serving them.

½ cup (125ml) urd dal
½ cup long-grain rice
1 teaspoon salt
2 teaspoons grated fresh ginger
Pinch of cayenne
2 tablespoons finely chopped chives
1 tablespoon grated or minced onion (optional)
Ghee or oil

Soak the dal and the rice separately for 3 hours. Drain very well and grind separately in a blender to a smooth paste. Mix the two pastes together, add the salt and leave overnight.

Add the ginger, cayenne, chives and onions. If the batter seems too thick it can be thinned with a little cold water.

Heat a heavy crêpe or omelette pan and brush with ghee or oil. Pour in 2 to 3 tablespoons of batter and cook for 2 minutes each side, turning carefully with a spatula. Keep warm, covered with a cloth, while you fry the remaining dosas. Serve as a bread.

Stuffed dosas: Prepare as above but before turning the dosa, fill with 1 tablespoon of vegetable filling made from mashed potato mixed with lemon juice, chopped parsley, chopped onion, a little ground coriander and garam masala (p. 68) to taste. Fold the dosa and keep warm, covered, while you prepare the rest.

Channna Dal Fritters
India

1 cup (250ml) channa dal
½ teaspoon salt
1 tablespoon minced or grated fresh ginger
2 teaspoons roasted and freshly ground cumin
½ beaten egg
1 tablespoon finely chopped parsley
Oil (for deep frying)

Cook the channa dal for 1 hour in lightly salted water or until soft. Drain and sieve to a smooth purée.

Mix in the ginger, salt and cumin and beat in the egg and parsley.

Heat the oil to 350°F (180°C) and drop table-spoonfuls of the batter into the oil. Fry until deep golden and drain well before serving.

Dahi Vadda
India

½ cup (125ml) urd dal, soaked for 3 hours
1 teaspoon minced or grated fresh ginger
1 teaspoon grated fresh coconut or unsweetened shredded coconut
2 tablespoons chopped parsley
½ teaspoon salt
½ teaspoon ground cumin
Oil (for deep frying)
Raita:
1 cup (250ml) plain yoghurt, chilled
1 teaspoon ground cumin
Pinch of salt
Garnish:
1 tablespoon chopped parsley
Pinch of cayenne pepper

Drain the dal and grind in a blender until fairly smooth. Add the ginger, coconut, parsley, salt and cumin. Grind again to a smooth paste.

Heat 2 inches (5cm) oil to 350°F (180°C) and drop tablespoonfuls of the batter into the oil – only fry about four at a time as they puff up during cooking. When golden, after 2 or 3 minutes, remove and drain on paper towels.

To make the raita, combine the yoghurt with the cumin and salt. Stir the lentil puffs gently into the yoghurt and sprinkle with cayenne and parsley.

Ta'amia   *Egypt*

These delicious spiced fritters make marvellous party food, and can be served with salads or flat pitta bread. They are a traditional Egyptian dish and in this form are always made with ful nabed. The Israeli Falafel, although often identical to Ta'amia, is just as often made with chick-peas.

1 cup (250ml) ful nabed, soaked and skinned
1 small onion or shallot, chopped
1 clove garlic, crushed
2 tablespoons chopped parsley
1 teaspoon ground cumin
Pinch of baking powder
½ teaspoon ground coriander
Pinch of cayenne pepper
¼ teaspoon salt
Oil (for deep frying)

Crush the beans coarsely and mix with all the other ingredients. Grind to a smooth paste, then leave for 2 hours in a covered bowl.

Heat 2 inches (5cm) oil to 350°F (180°C). Shape the paste into small, flat cakes and deep fry, a few at a time, until deeply golden. Drain well and serve hot.

Falafel: Proceed as above, substituting soaked and skinned chick-peas for the ful nabed. Flavour in the same way, adding a little lemon juice to taste. Sometimes soaked and dried bulgur (cracked wheat) or a table spoon of dry white breadcrumbs is added. Grind to a paste, shape as above and deep fry.

Acarajé   *Brazil*

These shrimp fritters, which are African in origin (as is much of the food on the east coast of Brazil), are normally cooked in dendê oil. This orange-coloured palm oil gives a distinctive colour and taste to foods cooked in it. Although recipes usually suggest the use of liquid annatto to colour ordinary cooking oil instead, I find that a pinch of paprika and turmeric stirred into the oil serves this purpose just as well.

Acarajé are normally served with a fiery chilli sauce (p. 74).

1 cup (250ml) black eyed peas, soaked overnight
½ small onion, finely chopped
¼ pound (100g) cooked shelled shrimp, roughly chopped
Salt and pepper
Oil (for deep frying), coloured with ½ teaspoon paprika and ¼ teaspoon turmeric

Drain the beans and remove their skins. Grind the beans to a paste in a mortar or blender. Stir in the onion and shrimp, seasoning with salt and pepper.

Heat the oil to 350°F (180°C). Drop in a few tablespoonfuls of the paste – about five at a time. Fry for 3 minutes on each side and drain well on kitchen paper. Keep hot while you fry the remainder of the paste.

Deep fried lentil cakes – good as a snack or with a curry.

1 cup (250ml) channa dal or split yellow peas, soaked overnight
1 onion, very finely chopped
1 inch (2¼cm) fresh ginger, peeled and finely chopped
1 small fresh green chilli, seeded and chopped, or ⅛–½ teaspoon powdered chillis (to taste)
Salt
Oil

Drain the dal and grind in a mortar or a blender to a smooth paste. Stir in the onion, ginger, chilli and salt to taste. Form into small flat cakes (or drop into the oil in tablespoonfuls– not so authentic but quicker and easier).

Heat the oil to 375°F (190°C) and fry for 3 or 4 minutes until golden brown. Drain well and serve hot.

Fried pastries filled with a spiced lentil purée, these are rather like Samosas and are excellent either as snacks or with an Indian meal. Puff pastry can be used instead of the dough given here.

Filling:
2 tablespoons oil
1 small onion, very finely chopped
1 clove garlic, finely chopped
½ teaspoon chilli powder or cayenne pepper
½ teaspoon ground ginger
1 teaspoon ground coriander
½ cup (125ml) masur dal, soaked for 3 hours
Dough:
1 tablespoon ghee or oil
1½ cups (375ml) all-purpose flour
½ teaspoon salt
⅓ cup (75ml) water or milk and water
Oil (for deep frying)

First, make the filling: heat the oil and fry the onion and garlic until soft but not coloured. Stir in the spices and cook gently for 2 minutes. Dry the lentils thoroughly and add to the pan, stirring. Barely cover with water and bring to the boil. Add the salt and simmer for 45 minutes until very soft and thick. Cool to room temperature.

To make the dough, rub the ghee into the flour and salt. Pour in the water and knead quickly to a soft dough, adding a little more water if too dry. Knead for 5 minutes and leave covered with a damp cloth until required.

Roll the dough out very thinly and cut into 3 inch (7½cm) circles. Damp the edges and place 1 teaspoonful of filling on each. Fold and seal, like miniature pasties – they should be crescent-shaped. Heat the oil to 375°F (190°C) and deep fry until golden brown. Drain and serve hot.

Dips and Spreads

Hummus bi Tahina   *North Africa*

This classic Arab dish has become so well known outside the countries of its origin that it hardly needs introduction. Although chick-peas are traditionally used, any other bean or lentil would be suitable with the tahina dressing.

1 cup (250ml) chick-peas, soaked
Juice of 1 lemon
⅔ cup (150ml) tahina paste
3 cloves garlic, crushed
1 teaspoon salt
1 tablespoon oil
1 teaspoon paprika
Garnish:
Finely chopped parsley
To serve:
Flat pitta bread

Cook the chick-peas for 1 hour. Drain, reserving a little of the cooking liquid.

Blend the chick-peas, garlic and a little of the cooking liquid to a smooth paste. Beat in the lemon juice, salt and tahina paste until smooth. Pour into a shallow bowl.

Combine the paprika and oil and trickle over the salad. Sprinkle with chopped parsley.

Lentil Pâté

A simple spreading paste which is good on toast, flat bread or in sandwiches with tomatoes and chopped celery. Other flavourings, such as tomato paste, tahina and soy sauce can be added to the paste to ring the changes.

It will keep for 3 or 4 days in a refrigerator if covered.

1 onion, very finely chopped
1 tablespoon oil
1 cup (250ml) brown or green lentils
1 clove garlic, crushed (optional)
¼ teaspoon dried thyme
¼ teaspoon dried marjoram
Salt and pepper
1 teaspoon anchovy paste (optional)
Melted butter

Fry the onion in the oil until golden. Add the lentils, garlic and herbs, cover with water and cook for about 40 minutes until soft.

Drain well and purée. Season and add the flavouring.

Put into small glass jars or earthenware pots and when cold, pour over enough melted butter to form a ½ inch (1¼cm) seal. When the butter is cold and solid, cover with foil or plastic film and store in the refrigerator.

Pot Likker
Sweden

More a custom than a dish, this consists of 1 pound (225g) yellow split peas cooked with a ham bone for 4 hours in 6 cups (1½ litres) water and lightly seasoned with ginger. It is brought to the table on New Year's Eve and everyone dips in a piece of bread for luck.

Aduki and Cream Cheese Paste

1½ cups (375ml) sprouted aduki beans, chopped
½ cup (125ml) cream cheese, ricotta or sieved
    cottage cheese
2 tablespoons chives, chopped
2 teaspoons oil
Salt
Pinch of cayenne pepper

Mix together the ingredients and use as a sandwich filler, on flat breads or to stuff tomatoes or green peppers which are to be eaten raw.

Ful Nabed Purée
Middle East

A beautiful, earthy salad which looks as good as it tastes. Serve as a dip with flat pitta bread and a bowl of olives.

1 onion, chopped
3 tablespoons olive oil
1 cup (125ml) ful nabed, soaked
Juice of 1 lemon
Salt and pepper
Dressing:
1 tablespoon olive oil
Juice of ½ lemon
1 teaspoon paprika
Garnish:
Finely chopped parsley

Cook the onion very gently in the oil, without letting it colour. Cover with 1¾ cups (425ml) water, add the beans and bring to the boil. Lower the heat and simmer for 1½ hours or until the beans are very soft. Purée to a smooth, creamy paste. Season with the lemon juice and salt and pepper to taste. Pour into a dish and leave to cool.

Combine the oil, lemon juice and paprika and trickle it over the cold purée. Sprinkle with the parsley.

Purée of Chick-peas   *Middle East*

Variations of this simple but deliciously earthy dish are popular in most Middle Eastern countries. The quantity of chick-peas used depends on whether it is served alone or to accompany meat. The garlic and yoghurt topping can be omitted, and the purée simply served with a dressing of oil coloured with paprika.

1½ cups (375ml) chick-peas, soaked
4–6 cloves garlic, according to taste
1 teaspoon salt
1–2 tablespoons olive oil
4–5 tablespoons plain yoghurt
½ teaspoon dried mint (optional)
1 teaspoon oil
½–1 teaspoon paprika
To serve:
Pitta bread

Cook the chick-peas until soft, about 1 to 1½ hours, depending on their age. Drain and reserve the liquid.

Crush 3 to 5 cloves of garlic with the salt, add to the chick-peas and pound to a paste or purée in a blender. Add the olive oil, beating well, and pour into a warmed dish.

Crush the one remaining garlic clove and mix with the yoghurt and mint, if used. Pour over the chick-pea purée. Mix the oil with the paprika and trickle over the yoghurt. Serve hot.

Stuffing for Baked Vegetables

This Middle Eastern stuffing of rice, chick-peas and tomatoes can be used for eggplant, onions, green or red peppers, tomatoes, zucchini or squash. It is best served cold and can be varied by using cumin, cinnamon or coriander. This recipe makes enough for 4 eggplants or peppers, or 6 zucchini, tomatoes or onions, depending on their size. They are best if left overnight, but do not serve chilled.

¼ cup (50ml) chick-peas, cooked and coarsely crushed
¼ cup (50ml) rice, uncooked
1 small onion, chopped
1 large tomato, peeled and chopped
Salt and pepper
Pinch of ground allspice
Few drops of lemon juice

Preheat the oven to 350°F (180°C). Mix all the ingredients together and fill the prepared vegetables. Place in an ovenproof dish and cover and bake for 30 minutes, or until the vegetables are tender. Allow to cool and serve.

Sweetmeats and Specialities

Dainagon no Amani   *Japan*

A sweet dish of aduki beans. These little red beans, which have a nutty flavour, are used extensively in Japan for sweet dishes and confections. Begin the day before.

⅔ cup (150ml) aduki beans, soaked
½ cup (125ml) sugar
Pinch of salt

Cook the beans for 1½ hours or until tender and most of the water absorbed. Add the sugar and salt and cook for another 20 minutes, stirring occasionally.

Transfer to a bowl and leave until the next day. Drain and serve.

Pakah   *Thailand*

These are also good if you add an equal quantity of shelled, unpeeled peanuts to the frying basket.

1 cup (250ml) lima or Great Northern beans, soaked overnight
Oil (for deep frying)
Salt

Dry the beans thoroughly, using a clean teatowel or some paper towels. Heat the oil to 375°F (190°C). Deep fry the beans, using a mesh basket if possible, until deep brown. Drain well and serve sprinkled with salt.

Sekihan   *Japan*

This and the following dish, Mizuyokan, are typical Japanese ways of using aduki beans. Sekihan is traditionally served on festive occasions, particularly at weddings. The rice should be mochi gome, Japanese sweet rice – if using short-grain rice add a little sugar to taste. Although Sekihan is sometimes served as a sweet course, the overall taste should be bland rather than sweet. It is often served with fish and can be eaten cold.

I must confess that I prefer aduki beans plain boiled with butter or in a salad, but have included these recipes for interest as they may appeal to those with a sweet tooth. Start a day in advance.

½ cup (125ml) aduki beans, soaked
¾ cup (175ml) sweet rice, washed
1 teaspoon sesame seeds, preferably black
Pinch of salt

Cook the beans in their soaking water for 30 minutes. Drain and reserve the liquid. Soak the rice in the bean liquid overnight to colour the rice. Drain, discarding the liquid, mix the rice with the beans and steam the rice and beans for 45 minutes in a fine mesh steamer or colander over boiling water. (If using short-grained rice, sweeten slightly to taste).

Dry fry the sesame seeds for 1 minute in a hot dry pan, stir in the salt and set aside. Serve the Sekihan hot, sprinkled with the sesame seeds.

Mizuyokan   *Japan*

1 cup (250ml) cooked aduki beans
2 tablespoons gelatine
½ cup (125ml) sugar
¾ cup (175ml) water

Cook the beans for an hour until very soft. Drain and purée to a very smooth paste. Drain again (wring out in a clean teatowel for the best result).

Dissolve the gelatine and sugar in the water over a medium heat. Bring to the boil and add the bean paste. Boil for 1 minute, stirring constantly. Pour into lightly greased shallow tins, cool and refrigerate. Serve cut into lozenge shapes, squares or triangles.

Index

GREEN BEANS ALMONDINE
Printed from COOKS.COM

1 (9 oz.) pkg. French cut green beans
1/3 c. boiling water
1/2 tsp. salt
1 1/2 tbsp. butter
1 tbsp. slivered almonds

Cook green beans in salted boiling water in a small covered saucepan, separating beans with a fork, 8 minutes, until tender-crisp. Heat butter in a small pan; saute almonds in butter until golden brown and butter is slightly browned. Drain beans and toss with almond mixture. Yield: 4 (1/2 cup) servings. Each serving containers: 2 g protein, 5 g fat, 65 calories, 197 mg sodium

7) Cook on a low heat for approximately 6-7 minutes. Drain off excess butter.

8) Add crushed almonds and toss beans. Lightly salt and serve.

Serves 8-10.

Submitted by: Patrick

String Bns. Almondine p. 7

Valerie Turvey

Bean Feast is Valerie Turvey's first book, but her experience as a cook goes back many years: as an art student she spent several vacations working in the kitchens of hotels and restaurants in Devon and Cornwall, and it was during this time that she gained her knowledge of all the basic cooking skills. Since her marriage in 1968 to Alan Turvey, a lecturer in graphics, she has continued to add to her collection of recipes and is now working on other cookery book ideas.

Valerie trained at the Plymouth College of Art and Design, where she obtained a Special Level Diploma in Illustration. After this she attended Birmingham College of Art and gained her teaching diploma in 1966. She was Head of the Art Departments at Swanage and Plympton Grammar Schools, and since her marriage has continued as a part-time teacher and freelance illustrator.